HOPE FOR TOMORROW:

JESUS AND TECHNOLOGY

Justin Tomkins

HOPE FOR TOMORROW

Jesus and Technology

First published in Great Britain 2021

A Sunnyside Books Publication

A CIP record for this book is available from the British Library.

ISBN 978-0-9576718-1-2

To Tuck & Olivia

with my love

ACKNOWLEDGEMENTS

This book both took form and drew to a close at Lee Abbey, the Christian community and conference centre in North Devon where Delana, now my wife, and I met back in 2002. These chapters are rooted in talks I gave within a Lee Abbey house-party in March 2019, and they started to draw together during a holiday visit in August 2021. I am grateful to Lee Abbey for the opportunity to lead those sessions back in 2019, and for providing a context in which I've experienced God at work in so many life-giving ways over these past decades.

Whilst the book took form at Lee Abbey, it originated earlier within my time at St Mary's Church, Longfleet (SML) in Poole. It was such a great joy and privilege to belong and work within that church family over a number of years and that sense of belonging continues, even if now from a distance. Thank you to Andy, to Mark, and to all at SML for your colleagueship and friendship and your various parts in shaping such a wonderful community within which this book began to grow. Thank you too to Josh and to all involved in 'Faith, Technology and Tomorrow'.

As this book began in SML, so it's grown and developed to a point when it's finally coming to birth within the Yeldham churches and the Hedinghams and Upper Colne Benefice (HUCB). Thank you to Liz and to all at HUCB for your welcome and your partnership in the local churches here, not least for your love and community through what's been such an unusual past eighteen months for us and for the world.

I am so grateful to family and friends for all your support and encouragement which has helped enable this project to come to completion, particularly those who have read and commented on drafts of this text including Andy, Delana, Emma, Lin, Lorna, Martha, Pete and Sally. However many flaws remain, there would have been far more were it not for your support. Mum, as well as your help in that way, your hospitality around the time of those Lee Abbey talks was such a great gift in making all this possible. Thank you.

Tuck and Olivia, we've spoken about this book together for a large part of your lives and now it's finally coming together! Olivia, thank you for your wonderful patience in waiting for a book upon which you can appear on the cover photo! For both of you, I pray that this book is one little bit of God's work equipping your generation as well as mine to be part of helping shape a hope-filled tomorrow. Delana, as with so very much in my life, this wouldn't have been possible without you. Thank you so much.

CONTENTS

INTRODUCTION

You may remember the 2013 advert for the French manufacturer *Le Trèfle*.[1] A husband teases his wife, Emma, for her old-fashioned reliance on paper rather than his technological alternatives. She uses paper-based worksheets, sticky notes, a puzzle book, a boarding pass and a novel. He has electronic versions of them all: all of them, that is, until he finds himself in the bathroom with the toilet paper gone. The advert ends with his look of shocked recognition when she passes him his electronic notepad under the bathroom door. On its display is a photo of a roll of toilet paper![2]

Some of us will relate more to the wife in that commercial and others to the husband. We may be early adopters of emerging technologies, quick to embrace new gadgets and to sign up for innovative platforms online. Others of us will be less keen to replace our tried and tested ways of going about our daily lives. We like the feel of our book as we read, we rely on our own map-reading ability for navigation and we wait for others to test out new devices. From Elon Musk and Jeff Bezos to Amish farmers in Pennsylvania, from toddlers born into a world of smartphones and *Netflix* to pensioners without email access, from impoverished young people in the Two-Thirds World who are hungry but on social media to Silicon Valley youth whose tech-rich parents limit their screen time, each of us will find ourselves somewhere on a matrix of technological engagement.

The preconceptions we bring to technology, both those we are conscious of and those we are not, will be shaped by our experiences as well as our character. Ray Kurzweil is an inventor and technologist

1

who went on to join *Google*. Back in 2007, his book *The Singularity is Near*[3] had sat unread on my bookshelf for a number of months. Then, that summer, I started to read. I was captivated. I remember an August evening, sharing Thai green curry with family and friends, and discussing his ideas on technology. Those ideas were, and are, both appalling and fascinating to me at once. I have no attraction for the world he paints in which we might choose to reject our own bodies and to upload ourselves onto virtual or physical replacements.[4] Yet I am convinced that he has an amazing ability to imagine and to describe the consequences of increasing growth and development in technology.

How do we assess our own responses to technology? Does the fact that I am not attracted to the world which Kurzweil paints make me a Luddite?[5] Is it possible that I am onto something when I feel appalled by his vision? Or does that sense simply represent my failure to embrace the new and the innovative? Is that emotion something I need to seek to overcome or to cherish?

The idea that we can reach beyond our own subjective feelings to a shared objective truth has been challenged for many decades now.[6] We may have learned to speak of being shaped by different stories. My thoughts and experiences are affected by my age, gender, ethnicity, income, education, and so on. I am enriched when I draw upon the perspectives of others and when I enter into community with those who are different. A male perspective is enriched by a female one. The young can be enlightened by the old, and *vice versa*.

We may have been taught to be suspicious of overarching stories. Certainly, if I see the world only through the lens of the white, or the male, or the middle-aged, I experience a distorted reality. Yet, as a scientist, I am not prepared to let go of the idea of truth. I can acknowledge that my own perspective is limited without losing my conviction that absolutes exist. Our technological context depends upon a shared understanding of such truths as the speed of light, the periodic table of the elements, and the structure of DNA.

We engage with technology from a perspective shaped by stories. I believe that some of those stories are more truthful than others. If my story excludes others, it needs to be stretched. If my story is not open to scientific truths, it needs to be stretched. So too, if my story is not consistent with the deeper realities of our world, it needs to be stretched.

As we explore our technological world, we will come across stories of those who wish to live to be a thousand years old.[7] We will encounter visions of the human race separating out into different species as parents seek to choose the genetic nature of their children.[8] We will meet those who seek to gain immortality through uploading their very self into a computer system, merging their own identity with technology.[9]

We will have feelings about these stories. Those feelings may simply represent prejudice, or lack of imagination, misunderstanding or selfishness. But they may not. Some of our reactions may come from spotting a vision which is built upon a story which is not truthful, or a story which is only for the benefit of some. Even if, at the level of intuition, we have not yet worked out what the emotion is telling us, some of those feelings are worth hanging onto, and analysing.

All of which means that, as much as ever, we need a way of testing stories and of analysing emotions. This book is written with the conviction that it is only in the person of Jesus Christ that we can begin to do that. I recognise that sentence will be a disappointment to some, maybe even a reason to cease reading any further. Yet all of us have a vantage point from which to engage and this conviction is mine. What's more, Jesus claimed that He is the Truth.[10] The Bible affirms that He made the world and all that is in it.[11] If those statements are correct, then what more truthful vantage point is there from which to engage with our contemporary world?

Let's slow down a moment though. What does it even mean to assert that stories can be tested and emotions analysed 'in the person of Jesus Christ'? What possible relevance can a Middle Eastern figure from two thousand years ago have for the world of 5G, *Zoom* and online shopping? Along with billions of others around the globe, I believe that Jesus of Nazareth was more than a famous carpenter from Roman times. I believe that he was a real human being but he was more. He was a good person, and a profound teacher, but he was more. My life and this book are built on the faith that Jesus Christ is God who made the universe and who became human out of personal love for you and for me.

That's all very well for me to say. Believing six impossible things before breakfast doesn't make them true. What evidence is there for this belief? I can't offer proof. If there were proof of God's existence then humanity would not have engaged in millennia of religious disputes. Yet there is evidence: evidence from the world, from the Bible and even evidence from within ourselves.

The fact that scientists can study our world witnesses to its order and predictability. We live in a world of dependable scientific laws. Apples reliably fall from trees. Even quantum events, with their inherent uncertainties, can be understood. It may not be easy to comprehend our universe but it is nonetheless intelligible. This intelligibility points to the faithfulness of a Creator God.[12]

4

As our world provides glimpses of the character of the One who made it, so too does the book which He gave us. The Bible tells us about ourselves and about God. The Bible offers a view of the world which rings true for me and for countless others, those alive today and those who have gone before.

That ring of truth is a reminder of the evidence for reality which we find within us. As we reflect upon our inner thoughts and feelings, we may be able to identify different voices and influences. Our hunger and tiredness will compete for attention with more profound ideas. We may be aware of the temptation to take an easy path even though we know of the pain that route may cause to others. Yet we may also recognise a voice calling out the best from within, pointing us to hope and to serve, a voice which calls us to love and which, as we listen, causes a sense of peace and of life to bubble up within us. Jesus promised His followers the gift of One with such a voice, One whom He called the Spirit of truth, the Holy Spirit.[13]

I had the privilege of working for a few months in a chemistry lab in Jerusalem. It was Spring and one of the highlights of my time there was celebrating Passover and Easter in Israel. On the Thursday evening, three days before Easter, I was part of a church service in the old city of Jerusalem. After the service itself was finished, I joined others from the congregation as, following Jesus' footsteps, we left the city and walked through the Kidron Valley to the Mount of Olives. Once we were there, in a garden, looking back towards the city, we stood in silence, praying, and reflecting upon the events of that first Easter.

What struck me was that I was cold! And not only cold, but surprised, astonished even, to be cold.

I recognised that, from childhood associations, my impression of Israel was of a hot country. Yet, here I was feeling cold! My brain could understand that Israel could be cold, particularly in the evening. Yet in a deep and unconscious way, maybe shaped by memories of Sunday School images, I had nonetheless pictured Jesus walking about in a land in which it was always hot. As the memory of a Bible passage, of a friend of Jesus, almost two thousand years before, standing by a fire to warm his hands[14] came to my mind, the Bible and the Holy Spirit helped to correct the distorted mental image which my experience had revealed. It was one small step in the renewing of my mind[15] and the recognition that biblical truth speaks to everyday embodied life.

My hope is that reflecting, within this book, on our technological context in the light of the Bible will bring similar every day, down-to-earth insights.

The overarching story of the Bible involves a number of core elements. One way of expressing those is in terms of the Creation of the world and its Fall; Israel and the law; the birth, death and resurrection of Jesus; the Church; and New Creation.[16] Those five elements, though in a slightly different order and combination, will shape the five pairs of chapters which comprise this book. Within each pair, the first chapter will identify biblical truths and the second will seek to use those as a means of interacting with our contemporary world. In this way, I hope to engage with technology as truthfully as I can, through seeking to hold various stories of technological visions and innovations alongside biblical truths of who Jesus is, who we are, and how He calls us to live.

So, this book grows out of questions such as these: How does our technological world connect to the Jerusalem of two thousand years ago? How do I worship Jesus in the world of Artificial Intelligence? How does the internet fit into a biblical overview moving from creation and Israel, through Jesus' birth, death and resurrection, to a new earth and a new heaven? What does it mean to be human? Am I a bad parent if I don't buy my son a smartphone?

The title of the book gives away the fact that I'd love this exploration to be grounded in hope not fear. That doesn't mean ignoring the seriousness of the various challenges we'll meet along the way. It does mean that because Jesus has already, through the cross and resurrection, secured the future, we can be confident in hope. It means trusting that even though we don't know what lies ahead in the short to medium term, we do know that a new heaven and a new earth await. We don't need to bring them about through our own efforts. Jesus has already secured that future for us. So, we're left with a question. How do we live in the meantime? This book is my attempt to explore what it means to follow Jesus in our technological world of today and tomorrow.

SUMMARY

1. Some of us are more inclined to engage with technology than others.

2. How do we assess the wisdom of our own response?

3. We need a truthful narrative to help us.

4. How about the biblical narrative of Jesus, who is Truth?

5. Evidence for the truth of who Jesus is comes from the world, the Bible and within ourselves.

6. The overarching story of the Bible may be thought of in terms of Creation; Israel and the Law; Jesus' birth, death and resurrection; the Church; and New Creation.

7. This is a hope-filled narrative.

8. This narrative will be used to shape this book.

QUESTIONS FOR REFLECTION

1. Who do you most identify with from the *Le Trèfle* commercial – the husband or the wife?

2. How do you experience the impact of technology upon your day-to-day life?

3. Can you think of an experience which has helped shape your feelings about technology?

4. What do you associate with the word 'truth'?

5. What hopes and fears do you have as you consider the future?

1   https://www.youtube.com/watch?v=-rf7khCkhGk Accessed 9th Sept. 2021.
2   I am grateful to Fi Perry for introducing me to this wonderful advert.
3   Ray Kurzweil, *The Singularity is Near*.
4   Ibid., particularly p299-317.
5   The association of someone who has issues with technology with the name 'Luddite' has its roots in the objections of nineteenth century English textile workers to the introduction of machinery which threatened their employment. Ned Ludd, their figurehead, may or may not have been a real person.
6   See, for example, *Truth is Stranger Than it Used to Be* by Middleton & Walsh which explores the subject of biblical faith in a postmodern age. Stanley Hauerwas is another of many theologians who engages with postmodernism. His book *A Community of Character* explores what it is to be story-formed community.
7   Chapter Six of this book engages with Aubrey de Grey's vision, as described in: Aubrey de Grey, *Ending Aging*.
8   Chapter Two of this book engages with Lee Silver's vision, as described in: Lee Silver, *Remaking Eden*.
9   Chapter Eight of this book particularly engages with Ray Kurzweil's vision, as described in: *The Singularity is Near* referred to above. Kurzweil's book *Danielle: Chronicles of a Superheroine* is also relevant to this area as are his two books co-written with physician Terry Grossman: *Fantastic Voyage* and *TRANSCEND*.
10 John 14:6: 'Jesus answered, "I am the way and the truth and the life. No one comes to the Father except through me."'
11 See for example, Colossians 1:16, 'For in him all things were created: things in heaven and on earth, visible and invisible, whether thrones or powers or rulers or authorities; all things have been created through him and for him.'
12 John Polkinghorne is one example of a scientist who has written about the intelligibility of the universe pointing to God's character. See, for example, *Reason and Reality*.
13 John 14:16-17: 'And I will ask the Father, and he will give you another advocate to help you and be with you for ever – the Spirit of truth. The world cannot accept him, because it neither sees him nor knows him. But you know him, for he lives with you and will be in you.'
14 John 18:18: 'It was cold, and the servants and officials stood around a fire they had made to keep warm. Peter also was standing with them, warming himself.'
15 Romans 12:2: 'Do not conform to the pattern of this world, but be transformed by the renewing of your mind. Then you will be able to test and approve what God's will is—his good, pleasing and perfect will.'
16 This biblical metanarrative as it is known is explored by Sam Wells in his book *Improvisation* as a means of engaging with what it means for us to improvise well as followers of Jesus at this point in biblical time.

1

IN THE BEGINNING

In the beginning God created the heavens and the earth.

Genesis 1:1

I wonder what gives you the greatest sense of awe. For some of us that will be looking up at the sky on a cloudless night, spotting stars, watching a meteor vaporise as it passes through the atmosphere or searching for a comet. For others of us it will be listening to recordings of whale song[1], or standing at the top of a mountain and looking down on the landscape below. It may be butterflies, or tigers, or considering which is the world's cutest baby animal.

I stood awestruck at the waterfalls in Iguazu, on the borders of Argentina, Brazil and Paraguay. I felt the spray on my skin. I not only heard the thunder of the cascading water but felt it in my feet and my legs as the earth itself reverberated with its power. I looked out and saw numerous other falls nearby and knew that this one, taking my breath away where I stood, was just one of around two hundred and seventy-five within the space of a few square miles.

That experience is etched in my memory, but as a research chemist by background it's the scale of the very small which has given me some of my most profound senses of awe. I'm fascinated by the world of chemical reactions in which colliding atoms make connections and so form new materials. I have loved to see microscopic images of cells and even molecules. Within my own research I was delighted by being able to visualise a strand of DNA connected to a protein as revealed using Atomic Force Microscopy. And I love the numbers involved.

Did you know that if you take a small glass of cold water, it will contain around 6 trillion, trillion water molecules?[2] That's 6,000,000,000,000,000,000,000,000. We don't know for sure how many stars there are in the universe, or grains of sand on all the beaches of the earth, but we do have estimates. There are between six and six hundred times more water molecules in that glass than the estimated stars in the universe.[3] There are close to a million times more water molecules in that glass than there are estimated grains of sand on all the beaches of our planet.[4]

The Bible affirms our natural sense of wonder. Repeatedly within its first chapter, God declares that His creation is good.[5] The Bible makes clear that we humans are creatures, made by God and that we too are well made.[6] This biblical description of a good creation affirms our feelings of wonder and beauty when we delight in a waterfall, a sunset, a rainbow, and even in one another.

The goodness of creation affirms the scientific enterprise of studying God's world. Because the physical stuff of creation is valuable and significant, because matter matters, it is well worth dedicating time, energy and even careers to understanding it. The scientist Werner Heisenberg, famed for his contributions to quantum theory, is reported to have stated that:

> The first gulp from the glass of natural sciences will turn you into an atheist, but at the bottom of the glass God is waiting for you.[7]

That seems to resonate with the biblical promise that the universe itself reveals the glory of God.[8] Whether we look to the very large or the very small we may glimpse God revealing Himself within His creation. As Jesus is our Creator God in human form, when we celebrate the goodness of creation, it is Jesus we honour.

Yet, we have only to check the news to know that not all is well in the world. A quick glance now at the BBC website, as I write, reveals a house explosion, the dangers of ultraviolet radiation, references to both terrorism and abuse, murder, clashes with police, racism, border disputes and the coronavirus pandemic.[9] And that's just on the homepage.

That same sense of something being wrong comes across in a particularly stark way in the form of the *Doomsday Clock*, a creation of the *Bulletin of Atomic Scientists*.[10] The *Doomsday Clock* was created in 1947 as a warning of how close humanity is to destroying the world with technologies of our own making. Each year, the *Bulletin of Atomic Scientists* assesses how close the hands of a symbolic clock need to be placed to midnight to reflect the level of threat faced by our world through the consequences of human action. Closer to midnight represents greater threat to the destruction of our planet and of our lives.

The clock was shown with its hands at seven minutes to midnight when it was first revealed in 1947.[11] In the years since, updated images have set out to answer two questions. Firstly, is the future of civilisation safer or at greater risk than it was last year? Secondly, is it safer or at greater risk today than during all the previous years, now spanning more than seven decades, within which these questions have been asked? The hands of that clock have moved backwards and forwards during the last seventy-four years, reflecting changes in nuclear concerns, which prompted the creation of the clock in the first place, and other global threats such as climate change, biotechnology, artificial intelligence and other emerging technological risks.

In January 2021 the clock's hands were judged to be at just 100 seconds from midnight, signalling as great a threat to humanity and to our planet as has been experienced at any other time since the clock's creation. When they first moved to that point one year earlier, the accompanying statement described that assessment as resulting from the 'two simultaneous existential dangers—nuclear war and climate change—that are compounded by a threat multiplier, cyber-enabled information warfare, that undercuts society's ability to respond'.[12]

I find these words deeply troubling. We live in a world of awe and wonder yet one in which not all is well.

The Bible is clear that God's good creation has been marred.[13] Theology names this 'the fall'. The tragedy of the fall emerges from the goodness of creation. One of the innumerable good gifts of creation is freedom. God gives us the choice of whether or not to follow Him. He allows us to be rebellious. Through freedom, human beings have chosen to do our own thing rather than to follow God. We live with the diseased consequences of that freely chosen rebellion. In Chapter Seven of this book, we will explore how Jesus has dealt with that problem once and for all. For now, we live in a world in which both beauty and brokenness are apparent and in which we wait for ultimate restoration.

A further creation truth revealed in the Bible is that, within creation, we humans are unique in being made in the image of God, and we thus have a distinct role and responsibility within the world.[14]

In ancient times in the Middle East, kings would set up statue representations of themselves in far off lands, to remind the locals of their rule and authority. Today flags are used to represent nations in a similar fashion. In the same way, all human beings have an innate role as a representative of God. As humans we are made in such a way that our very being is to point to God's goodness. How we live our lives will determine how effectively we fulfil that potential.

Many books have been written exploring what it means for human beings to be made in God's image.[15] The idea of representing God is one facet of the rich symbolism of the phrase. That symbolism points to the necessity of treating others with appropriate respect. It also reveals that as creatures made in the image of God, we too are creative, yet not as God. The expression 'playing God' taps into something of the fine line between being fully alive as a human being, expressing all the creativity which that involves, without seeking to be 'like God' and overstepping the mark into hubris.[16] Of course, there are many different opinions about where that line lies. Activities which for some will be an expression of our God-given creativity, will be for others a flagrant and disrespectful step too far.[17]

All of creation has immense value and significance as part of God's good creative work. Human beings have a particular significance and value through having been specifically chosen to image and to reflect God, our Creator.[18] Our very being points to Him.

In the next chapter, these biblical truths of creation will be used as a means of engaging with some of the technological issues we encounter in our contemporary world.

SUMMARY

1. Many of us experience creation with a sense of awe and wonder.

2. Biblical Truth One: Creation is good and Jesus is our Creator God in human form.

3. Biblical Truth Two: Humans are creatures.

4. Biblical Truth Three: Creation is marred as a consequence of the fall.

5. Biblical Truth Four: Humans are made in the image of God and so have a role as representatives of God.

QUESTIONS FOR REFLECTION

1. What gives you the greatest sense of awe?

2. How does the quote from Werner Heisenberg relate to your own experience?

3. Do you think the scientists who set the Doomsday Clock are right to place the hands as close to midnight in 2021 as at any point in the past 74 years? Why?

4. Are you more aware of the beauty or the brokenness of creation?

5. Can you think of an example of someone 'playing God'?

1   Louie Giglio creatively draws upon the wonder of both stars and whale song in his Passion Talk Series, *Symphony: I Lift my Hands* featuring the mashup of stars and whales.

2   Avogadro's number, named after the Italian scientist Amedeo Avogadro, gives the number of units in one mole of that substance, for example the number of water molecules in 1 mole of water. Avogadro's number is just over 6×10^{23} per mol, that's 6 followed by twenty-three zeros. One mole of water molecules has a mass of just over 18g, so a small glass of water weighing 180g contains around 6×10^{24} water molecules, 6 followed by twenty-four zeros.

3   This figure is based upon the European Space Agency's estimate of 10^{22} to 10^{24} stars in the universe: https://www.esa.int/Science_Exploration/Space_Science/Herschel/How_many_stars_are_there_in_the_Universe Accessed 9th Sept. 2021.

4   Robert Krulwich quotes a (very rough) figure of 7.5×10^{18} for the number of grains of sand on earth: https://www.npr.org/sections/krulwich/2012/09/17/161096233/which-is-greater-the-number-of-sand-grains-on-earth-or-stars-in-the-sky Accessed 9th Sept. 2021.

5   The first chapter of the Bible contains the repeated phrase 'And God saw that it was good'. In Genesis 1:31a, at the completion of Creation, we read: 'God saw all that he had made, and it was very good.'

6   Genesis 1:24-30; Genesis 2:4-25.

7   David Hutchings & Tom McLeish, *Let There be Science*, p87, quoting Ulrich Hildebrand.

8   Romans 1:20: 'For since the creation of the world God's invisible qualities—his eternal power and divine nature—have been clearly seen, being understood from what has been made, so that people are without excuse.'

9   The day in question was 25th June 2020.

10 https://thebulletin.org/doomsday-clock/ Accessed 9th Sept. 2021.

11 1947 holds a special meaning for me as it was the year in which both my parents were born. My father would not have survived had it not been for recent developments in blood transfusions which enabled his Rhesus disease to be treated shortly after birth.

12 https://thebulletin.org/wp-content/uploads/2020/01/2020-Doomsday-Clock-statement.pdf Accessed 9th Sept. 2021, p3.

13 Genesis 3.

14 The phrase 'image of God' appears in the Old Testament in Genesis 1:26-27, Genesis 5:1 and Genesis 9:6 and is referred to in a number of New Testament contexts.

15 *The Liberating Image* by Richard Middleton is one example of a book which explores the *imago dei* (image of God) in Genesis 1.

16   In *Respecting Life* p37-38 Neil Messer contrasts actions that conform to the *imago dei* (image of God) with diabolically-inspired attempts to be like God.

17   Co-creation is a theological name given to the idea of humans participating in God's ongoing work of creation. The subject of humans as co-creators is explored by a number of different writers within *Design and Destiny* edited by Ronald Cole-Turner.

18   As will be explored in Chapter Five, Jesus' incarnation as a human being adds further significance to the human form.

2

A LIFE OF AWE AND WONDER

One cannot help but be in awe when (one) contemplates the mysteries of eternity, of life, of the marvellous structure of reality.
Albert Einstein[1]

As described in the Introduction, the chapters in this book are paired and the second in each pair will seek to engage with our contemporary world. The previous chapter has already sought to articulate some biblical truths of creation. Those were:

1. Creation is good and Jesus is our Creator God in human form.

2. Humans are creatures.

3. Creation is marred as a consequence of the fall.

4. Humans are made in the image of God and so have a role as representatives of God.

It is now time to consider some stories of our contemporary world before seeking to reflect upon those stories in the light of the biblical truths from the previous chapter.

***

Whether or not our eyes are windows into our soul, they certainly, when working well, provide a doorway outward, through which to engage with the world around us. Yet that doorway out is increasingly mediated by a screen. During the time of Covid-19, screens have kept us in contact via *Zoom* and *FaceTime*. Through screens we have been introduced for the first time to new-born family members and through screens we have attended funerals from across oceans. Through screens we have ordered online shopping and looked in on government briefings as we have heard the latest statistics and been told about social restrictions. Through screens we have volunteered to serve, we have ordered testing kits, and we have confirmed vaccination appointments. Through screens our children have entered into virtual classrooms and have submitted assignments. Through screens we have played the soundtracks to our lockdown and escaped into movies and TV series. Through screens we have joined one another in online worship services and connected together for corporate prayer.

To condemn screen-use would be to deny the benefits is has brought, not just in these past months, but before that too. Yet to accept all the ways in which our use of screens is affecting human life, without reflection, would be to turn a blind eye to a major social influence of the early twenty-first century.

Susan Maushart is someone who was reflecting upon these issues years before lockdown. On 4th January 2009, Susan, journalist and mother of three teenagers, pulled the plug on her electricity supply to launch a six-month family 'Experiment'. Whilst she turned the power back on in their Australian home two weeks later, all screen-use within the house was off limits for the whole family until July. Furthermore, Susan gave up use of her cherished *iPhone*, even outside the home, throughout that time. The children acquiesced to the project through a combination of bribery and necessity, although the youngest of the teens did move out of the house to stay with her father for the first couple of months. It took the arrival of a new kitten to help lure her back home!

Before the start of the 'Experiment', Susan had become fascinated by the way that screen-use had accelerated throughout the world at the start of the twenty-first century. She was struck by figures, now more than ten years old, of American teenagers spending, on average, the equivalent of a full working day, seven days a week, in front of some form of screen.[2] The fact not only that more recent research shows that social media use has further increased since that time,[3] but that our own experience makes that finding obvious, indicates that the issue has only become more significant since Susan's initial interest.

Susan writes about the experience of her own family in the wonderfully entitled *The Winter of our Disconnect*, playing on the name of the 1961 novel by John Steinbeck.[4] She describes being inspired by the Australian microbiologist, Barry Marshall.[5] In 2005 Marshall won a Nobel prize for his discovery that stomach ulcers are caused by bacteria. Having initially been unable to persuade a drug company to test his theory with a clinical trial, Marshall swallowed some of the bacteria himself, promptly developed an ulcer, and so gained the foundational evidence required to continue his research. As Marshall had done before her, Susan decided to turn her own life into a 'Petri dish'.[6] Her family 'Experiment' would provide the opportunity to experience first-hand the significance of screens to contemporary life.

Susan set boundaries to enable the family to continue meeting commitments for work and school. Each family member remained

23

free to use screens outside the home, and some of the children spent far more time at friends' houses as a consequence! The family didn't find the six months easy, but they did find it fruitful.

One of Susan's two daughters caught up on sleep and as a result the rest of her life blossomed. She learned to cook, spent more time with her family, and had new-found energy with which to enjoy life.[7] As a result of not spending so much time playing online games, Susan's son became an accomplished jazz musician during the six months and, through doing so, made a whole group of new friends.[8] Susan's older daughter passed her driving test, completed two internships and had a magazine article published.[9]

Susan recognised that some of these accomplishments would have happened anyway, but she's in no doubt that the 'Experiment' helped. The family shared conversations and meal times with a frequency and quality that they hadn't previously enjoyed. They returned to screen-use at the end of six months but they returned changed, strengthened and with new perspectives.

The six-month 'Experiment' lived by Susan Maushart and her family demonstrates the fact that whatever happens on a global scale, each of us has the freedom to be discerning in our own engagement with technology.

***

Recognising creation as good might help us to reflect upon Susan Maushart's experience. She wrote repeatedly throughout her book about the inspiration of David Thoreau[10] and his commitment to reconnect with nature and wilderness. She describes the experience of noticing the details of her local area afresh when walking without having sounds played directly to her ears through an electronic headset.[11] We have already noted the impact of the 'Experiment' upon her children in terms of sleep, new friendships, and the time to develop new skills.

One tool with which to reflect upon our use of technology concerns the extent to which that behaviour either affirms or denies the goodness of creation. If our screen-use steals our sleep, takes our attention from people around us, and robs us of time and energy, then it deprives us of our ability to interact with the world. We might then challenge those patterns on the basis of the goodness of creation. If instead, our screens nurture and deepen healthy relationships with others, give us a greater sense of awe and wonder for the natural world, and help us use our time in a way that causes us to connect more deeply with creation, then we might affirm those behaviours on the same basis.

There are no easy ways to extrapolate from an appreciation of the goodness of creation to generic rules for using screens in our society today. What might help me might not help you. We can though, each search for patterns of screen-use that work for our own unique characters. We might amend those patterns over time. Six months later, in different circumstances, I might need to change my behaviour again and to adopt some new habits. As we have experienced during Covid lockdown, in one season it might be connecting with others online which is the priority. At another time, it may be more important to sacrifice some of the online to make more space for connecting in real life.

We might start to recognise patterns which feel life-giving. We might discover that it helps us to limit our reading and sending of emails, texts and social media messages to certain times and locations.[12] As we experiment, we might find it helpful to keep on reflecting on the question of whether our use of screens is helping or hindering us in connecting more deeply with the natural world and with other people.

The truth of the goodness of creation is not limited, of course, to the topic of screen-use. All other things being equal, it might lead us to affirm driving to a place of natural beauty inaccessible by other means, but might discourage us from using a car simply to avoid walking. It might lead us to affirm the building of a new telescope to study astronomy but to discourage doing so in a location that would threaten local wildlife. All other things being equal, it might lead us to affirm making a documentary on an endangered animal species but discourage doing so in such a way that increased tourism might threaten the very environment it sought to preserve.

All other things are not of course equal. The truth of the goodness of creation is just one element of the biblical revelation. Other elements too offer us tools for wise discernment in our technological context. We will come on to a further three in the remainder of this chapter. Nonetheless, reflecting upon how our technological decision-making either deprives or nurtures our appreciation of the goodness of creation is at least a place from which to begin.

***

Lee Silver is a molecular biologist from Princeton University. He argues that when it comes to thoughts of having children, the ideas of 'reproducing ourselves' and giving the resultant child 'the best possible life' are immensely influential.[13] Those aims which are embedded in very high numbers of potential parents across the globe can result in biological driving forces more powerful than any cultural pressure or government decree. When those common urges are allied to future developments in reproductive technologies, Silver argues that we will witness radical changes in human genetics.

Silver does not imagine a future world in which social programmes dictate that all children must be born in a particular way. He does not envisage a tomorrow in which political leaders decree that all parents must follow a single reproductive pathway. Simply by assuming that potential parents will make personal decisions about having a 'biological child'[14] and giving that child the best possible start in life, and that reproductive technologies will provide increasingly powerful tools for doing so, Silver describes a future of profound change.

This future is one in which 'natural' conception and childbirth continues. Nonetheless, alongside children born 'naturally', Silver imagines 'genetically-enriched' children born through the use of new reproductive technologies. As developing technologies provide successively powerful tools with which to implement reproductive choices, Silver describes generations of potential parents choosing different aspects of their child's make up. First it will be possible to select eye colour. Years later the possibilities of selecting for resistance to different infections and diseases will have increased. Further ahead still, children may be selected before birth not just with a potential to excel at sport but for a calculated excellence in a specific role within a particular sport.

Silver continues this thought experiment beyond the point where it will be impossible for any Natural child to compete with a genetically-enriched (GenRich) peer, to a time when:

> the GenRich class and the Natural class will become the GenRich humans and the Natural humans – entirely separate

27

species with no ability to cross-breed, and with as much romantic interest in each other as a current human would have for a chimpanzee.[15]

Lee Silver describes a future in which humanity will split into two different species, the naturals and the genetically-enriched. Whilst the naturals will be recognisable as what we understand as humans today, the genetically-enriched will find themselves on a new and different path. Further ahead, the genetically-enriched will split further, creating many different species as the descendants of humans spread out around the galaxy, each to thrive in an environment for which their particular genetics make them well-suited.[16]

**

The biblical description of humans created by God illuminates Lee Silver's vision of a future in which humanity separates into multiple different species. As will be explored in due course, the significance of the human form is not limited to our identity as creatures, yet this identity alone is enough to lead us to question any vision in which the future of the human form is threatened.

Humans are not here by accident. The Bible assures us we are here through the will of a Creator who looked upon all that He had made and declared it to be very good. If this is true then to risk losing the human form would be to threaten the existence of something of very great worth. To take Michelangelo's statue of David and to crush it into builder's sand would be a shocking re-purposing of a work of art but just a dim shadow of what it would be to risk the human form being lost through deliberate genetic conversion to a different design.

Such a plan would be to deny our identity as creatures and to seek to be a god. These aspirations are indeed alive and well within our society. Yuval Noah Harari has written a number of bestselling books on humanity past, present and future.[17] In the first chapter of *Homo Deus,* he explores 'A new human agenda', writing:

> ... at the dawn of the third millennium, humanity wakes up to an amazing realisation. Most people rarely think about it, but in the last few decades we have managed to rein in famine, plague and war. Of course, these problems have not been completely solved, but they have been transformed from incomprehensible and uncontrollable forces of nature into manageable challenges. We don't need to pray to any god or saint to rescue us from them. We know quite well what needs to be done in order to prevent famine, plague and war – and we usually succeed in doing it.[18]

He asks what projects will replace famine, plague and war at the top of the human agenda in the twenty-first century[19] and then, later, he writes:

> In seeking bliss and immortality humans are in fact trying to upgrade themselves into gods. Not just because these are divine qualities, but in order to overcome old age and misery humans will first have to acquire godlike control of their own biological substratum. If we ever have the power to engineer death and pain out of our system, that same power will probably be sufficient to engineer our system in almost any manner we like, and manipulate our organs, emotions and intelligence in myriad ways. You could buy yourself the strength of Hercules, the sensuality of Aphrodite, the wisdom of Athena or the madness of Dionysus if that is what you are

into. Up until now increasing human power relied mainly on upgrading our external tools. In the future it may rely more on upgrading the human body and mind, or on merging directly with our tools.[20]

He continues:

> We want the ability to re-engineer our bodies and minds in order, above all, to escape old age, death and misery, but once we have it, who knows what else we might do with such ability? So we may well think of the new human agenda as consisting really of only one project (with many branches): attaining divinity.[21]

Such aspirations can be challenged on the basis that they express a rejection of our human identity as creatures, made by a Creator God.

***

Maisy McAdam started to lose her sight at the age of sixteen due to a brain tumour.[22] Over the following few months, her vision was reduced to a small blurry circle in her right eye. She was registered blind as a consequence. Yet at the 2019 Hay Festival she was invited, live on stage, to try on a pair of *Give Vision* goggles for the first time. The goggles used magnifiers and Augmented Reality (AR) to amplify Maisy's remaining vision and to highlight outlines. To her own amazement and that of the audience, for the first time in six years, Maisy was able to read a book. Technology enabled Maisy to regain her sense of sight.

Other high-tech glasses offer the chance for those who can already see to add to that ability. Since the launch of *Google Glass*[23] in 2013, 'smart glasses', with or without prescription lenses, have offered users the chance to enhance natural vision. Options include Augmented Reality through full colour high-definition projections onto a lens and cameras giving others a chance to follow one's vision. Users can send and receive messages, access information online and make use of embedded microphones.[24] Some of the glasses look decidedly high tech but others appear much like any other pair of spectacles, so that no one else need even be aware of their functionality.

In October 2020, the BBC reported news of the US army testing the use of AR goggles for combat dogs.[25] Military dogs can seek out hazards, but they need guidance. These goggles are designed to create the possibility of handlers directing the dogs more effectively from a safe distance.

In current combat deployments, soldiers usually direct their animals with hand signals or laser pointers - both of which require the handler to be close by. But that need not be the case if the prototype AR goggles are widely adopted, the army said. Inside the goggles, the dogs can see a visual indicator that they can be trained to follow, directing them to a specific spot. The handler, meanwhile, can see what the dog sees through a remote video feed.[26]

The three devices described above - Maisy's goggles, smart glasses and the googles worn by the combat dogs - all rely on Augmented Reality. Yet this one technology is put to different uses in the different devices. Maisy's goggles restore sight which has been lost. The others seek to go beyond sight which already exists. We will return to this distinction in Chapter Four.

***

The fact that creation is marred means that our bodies need healing. As the biblical revelation of the goodness of creation affirms science, so too the biblical revelation of the fall affirms the importance of medicine. Facing the realities of life in a fallen world draws attention to the need for healing. Throughout history Christians have been at the forefront of meeting the needs of those who are ill. As Christianity has helped nurture the emergence of science, so too it has helped develop the field of medicine.

Jesus' healing work was a key aspect of his ministry as recorded in the Bible. The early church continued this focus. In later centuries abbeys and monasteries helped develop early medical care. During times of plague and disease, Christians were renowned for being prepared to put their own lives in danger in order to care for others. Today, it is a desire to follow Jesus which prompts large numbers of men and women to dedicate their lives as doctors, nurses and in other healthcare roles. As the Church continues to pray for healing for the sick, so too it works to bring about such healing through medicine.

A number of those present at that 2019 Hay Festival cried with joy when they witnessed Maisy McAdam reading for the first time in six years. There is a recognition deep within most of us that healing is to be celebrated.

Seeking to draw upon the goodness of healing quickly leads us, however, into complex territory. Augmented Reality returned Maisy's sight in an act of healing but was deployed for a different purpose in the goggles of the combat dogs. As noted above, we will return to this in Chapter Four. For now, we might at least conclude that when technology is used for healing the sick (as opposed to seeking to go beyond health), such a purpose can be affirmed on the basis that creation is marred and in need of such healing.

***

On the morning of Friday January 14th 2011, David Ferrucci knew that by the end of the day his whole team would have been judged on the basis of a television show. The journey towards that high-profile appraisal had started around six years earlier when *IBM* executives were searching for the next 'Grand Challenge'. *Deep Blue*'s victory over chess Grand Master Garry Kasparov in 1997 had cemented *IBM*'s reputation as a technological heavy-weight[27] but the company's leaders knew that business reputations require continual maintenance. The industry moves too quickly to allow even key competitors to bask for long in former glories.

Stephen Baker recounts the story of the *Jeopardy* contest in his 2011 book, *Final Jeopardy*.[28] Some say the original idea for the challenge came from an *IBM* researcher named Charles Lickel who was dining out one evening in autumn 2004 when he was struck by the fact that the restaurant suddenly emptied as diners left food untouched and filed into the bar to watch that evening's broadcast of the quiz show *Jeopardy*.[29] Whatever the roots of the project, by the summer of 2006 David Ferrucci had somewhat reluctantly accepted the leadership of a team building a machine to beat human champions of the popular American quiz show in which each answer is a question.[30] That machine would need to be able to respond to a clue such as 'Four-letter word for the iron fitting on the hoof of a horse, or a card-dealing box in a casino', with the correct response, 'What is a shoe?'

To win at *Jeopardy*, the *IBM* team needed to produce a system which could not only understand complex English language, and search information at speed, but which could also evaluate between different possible answers, whilst dealing with humour and word play.

There were numerous challenges *en route* as well as some failures in the high profile shows in which *Watson* competed against *Jeopardy* champions Ken Jennings and Brad Rutter. At one point, *Watson* failed to correctly provide the question 'What is chess?' to the clue, 'Garry Kasparov wrote the foreword for The Complete Hedgehog, about a defence in this game.' *Watson*'s predecessor, *Deep Blue*, might have been disappointed! But despite those road bumps, seven years after the project began, *Watson* went on to win the two-part final with a total of $77,147. Jennings finished a distant second, with $24,000, just ahead of Rutter with $21,600. Jennings responded with the poignant words, 'I, for one, welcome our new computer overlord.'[31]

***

How might a biblical understanding of creation help us to reflect upon the work of the *IBM* team in designing and producing a machine to become a quiz show champion? As creatures made in the image of the Creator God, we too are called to be creative. We have been enabled to imagine, to design, to craft and to develop our own mini-creations within the world. We are not God and we cannot create out of nothing, but within this world which He has created, rich in innate potential, we can create in turn. As we do so appropriately, we step deeper into our God-given identities. God delegated to Adam the task of working in the Garden of Eden, caring for it and naming the animals.[32] There are myriad ways in which He calls us to exercise our own gifts with creativity and imagination.

Watching a machine accomplish a task which no machine has succeeded at before, and which until this point had been the preserve of human endeavour may prompt an appropriate sense of wonder. If so, that wonder points back to the reality that God has created a world in which humans, created in His image, have been given so much potential for creativity that such machines are a product of that God-given gift.

Such creativity needs, of course, to be wielded responsibly. One apocalyptic scenario much-explored by technologists involves the universe being converted into one enormous paperclip-manufacturing system.[33] The idea is that if a sufficiently powerful AI is given a poorly expressed instruction, that AI might fulfil those instructions to the letter and so cause mass devastation. Imagine an AI instructed to produce paperclips which does so with such efficiency that not only is humanity itself wiped out in the process but that the resources of the universe are all turned to the single purpose of making yet more paperclips. As programmer and philosopher Eliezer Yudkowsky is quoted as saying: 'The AI does not hate you, nor does it love you, but you are made of atoms which it can use for something else'.[34]

It seems to me that however else we reflect upon the creation of IBM's Watson, that work expresses something of humans made in the image of God exercising their own God-given creativity. That's not to give carte blanche to any form of human creativity. It's not to affirm the creation of weapons of mass destruction. We may still wish to reflect upon the risks and benefits of developing artificial intelligence and the impact of Watson upon that quest.[35] We might choose to consider how other biblical truths further illuminate this project. Yet, as a starting point, whenever we encounter human creativity, whether technological in nature or not, we can at least affirm that rich creativity with which we have been endowed as beings made in the image of God.

***

No chapter on creation would be complete without at least acknowledging the pressing environmental challenges we face. Climate change is increasingly viewed as at least one of the most important issues which our global community needs to address.[36] Vast numbers of species currently teeter on the brink of extinction.[37] The health of our planet requires our urgent attention.

The focus of this book is upon technology rather than the environment. Nonetheless, we may find ourselves with questions at this point. Can technology solve our environmental problems? Will it make them worse? If the Bible can help us think about technology, can it help with the environment too? Will an environmental catastrophe wipe out technological progress and make the reflections of this book pointless? Will technology or the environment have a more significant impact upon the coming decades? What difference can I make anyway?

Appendix Two, *Technology, the Environment, and the overarching story of the Bible*, at the back of this book, offers a few signposts for further reflection on these issues.

***

As we continue to explore how biblical truths illuminate our technological context, we will identity further tools with which to discern the appropriateness of different uses for our human creativity. For now, let's summarise our engagement with the biblical truths of creation.

The biblical truth of the goodness of creation leads us to affirm that which nurtures our appropriate engagement with the natural world and one another. Our identity as creatures cautions us about any plans which might threaten our existence or express the desire to become like gods. The reality of the fall prompts us to work for healing and to celebrate appropriate uses of medicine. Our identity as creatures made in the image of God leads us to recognise that we have been created to be appropriately creative in turn.

What are appropriate boundaries around these various activities? Our exploration continues as we move on to explore how God's gift of the law offered His people Israel a means of understanding how to live well as creatures made in His image and identified boundaries for doing so.

SUMMARY

1. Susan Maushart has explored the topic of screen-use in our society by disconnecting from screens in her home for six months.

2. The biblical truth of the goodness of creation invites us to distinguish technologies on the basis of whether they nurture or deprive our appreciation of the goodness of creation.

3. Lee Silver imagines a future in which humanity will split into different species as a consequence of genetically engineering children.

4. Yuval Noah Harari presents a vison of the future in which humans seek to become 'gods' through the use of technology.

5. The biblical truth of humans as creatures challenges any technological vision in which the future of humanity might be threatened or in which humans might seek to become 'gods'.

6. Maisy McAdam, having become blind, was able to read for the first time in six years with the aid of *Give Vision* goggles.

7. The biblical truth of creation being marred affirms the healing use of technology, where healing is distinguished from attempts to go beyond health.

8. *IBM*'s *Watson* beat human champions at the quiz show *Jeopardy* in 2011.

9. The biblical truth of humans made in the image of God affirms appropriate human creativity.

10. The 'Paperclip Apocalypse' is a vision of the consequences of giving poor instructions to highly capable machines.

QUESTIONS FOR REFLECTION

1. Was it unfair of Susan Maushart to disconnect her children, even temporarily, from the online world in which their friends were spending so much time? Why?

2. Do you think Lee Silver's vision of humanity separating out into many different species is realistic? Why?

3. What does it mean to give a child the best possible start in life?

4. What makes a great sculpture more valuable than the material from which it is made?

5. How do you feel about Harari's idea that humans are seeking to upgrade themselves into gods?

6. What emotions do you imagine Maisy McAdam felt as she read words upon a page for the first time in six years?

7. How do you feel about combat dogs equipped with Augmented Reality goggles?

8. What might *IBM* pursue as its next 'Grand Challenge'?

1   Quoted by William Miller within a personal memoir in *Life* Magazine, 2 May 1955, p64, https://books.google.co.uk/books?id=dIYEAAAAMBAJ&printsec=frontcover#v=onepage&q&f=false Accessed 9th Sept. 2021.

2   Susan Maushart, *The Winter of our Disconnect*, p4. Maushart quotes research carried out by the Kaiser Family Foundation reporting that the average American teenager was spending 8.5 hours a day in some form of mass-mediated interaction.

3   https://datareportal.com/reports/digital-2020-july-global-statshot Accessed 9th Sept. 2021.

4   John Steinbeck, *The Winter of our Discontent*.

5   Susan Maushart, *The Winter of our Disconnect*, p7.

6   A Petri dish, named after German bacteriologist Julius Richard Petri, is a shallow lidded dish used by scientists to grow cells such as bacteria.

7   Susan Maushart, *The Winter of our Disconnect*, p252-253.

8   Ibid., p252.

9   Ibid., p253.

10   Henry David Thoreau, 1817-1862, was an American naturalist who wrote about his experience of living simply, by Walden Pond in Massachusetts, within his book *Walden: On Life in the Woods*.

11   Susan Maushart, *The Winter of our Disconnect*, p130-131.

12   The technology columnist, Kevin Roose, has written about 'demoting our devices' within his 2021 book, *Futureproof*. He describes his own experience of being supported by Catherine Price, author of *How to break up with your phone*. In her 2017 book, *Left to their own Devices?*, Katherine Hill explores how parents and grandparents might help children tackle issues of screen use.

13   Lee Silver, *Remaking Eden*, p1-13, 308.

14   Ibid., p308.

15   Ibid., p8.

16   Ibid., p286-290.

17   Yuval Noah Harari: *Sapiens*; *Homo Deus*; *21 Lessons for the 21st Century*.

18   Yuval Noah Harari, *Homo Deus*, p2.

19   Ibid., p2.

20   Ibid., p49-50.

21   Ibid., p53-54.

22   https://www.bbc.co.uk/news/av/technology-48501600/goggles-give-back-sight-to-maisy-so-she-can-read-again Accessed 9th Sept. 2021.

23   https://www.google.com/glass/start/ Accessed 9th Sept. 2021.

24   https://www.youtube.com/watch?v=5dsVwXzOIjk Accessed 9th Sept. 2021.

25   https://www.bbc.co.uk/news/technology-54465361 Accessed 9th Sept. 2021.

26   https://www.bbc.co.uk/news/technology-54465361 Accessed 9th Sept. 2021.

[27] *IBM's* chess-playing computer, *Deep Blue*, famously beat Garry Kasparov, the Chess Grand Master at the time, in a series of high-profile games in 1997. Garry Kasparov reflected back upon this experience twenty years later within his book *Deep Thinking*.

[28] Stephen Baker, *Final Jeopardy*.

[29] Ibid., p20.

[30] Ibid., p41.

[31] Ibid., p251.

[32] Genesis 2:15, 19-20a.

[33] Tom Chivers explores this scenario in his book *The AI does not hate you*, p60-61.

[34] Ibid., p94.

[35] Numerous technologists have warned of the possible dangers posed by AI. Bill Joy's article, *Why the Future doesn't need us*, published in issue 8.04 of *Wired* magazine in the year 2000 is a key example. Another is an open letter from the *Future of Life Institute*, published in early 2015 and signed by Stephen Hawking and Elon Musk amongst others. See Max Tegmark, *Life 3.0*, p33-37.

[36] One hundred and ninety-seven countries have now signed up to the Paris Climate Agreement. That is nearly every nation on earth. COP26, the UN Climate Change Conference UK 2021, is expected to include participants from every one of those 197 countries. Preparatory material for the conference describes climate change as 'the greatest risk facing us all'. https://ukcop26.org/wp-content/uploads/2021/07/COP26-Explained.pdf Accessed 9th Sept. 2021, p9.

[37] The World Wildlife Fund Living Planet Report 2020 reported a 68% drop in wildlife populations between 1970 and 2016. https://www.zsl.org/sites/default/files/LPR%202020%20Full%20report.pdf Accessed 9th Sept. 2021.

3

ISRAEL AND THE LAW

The LORD said to Moses, 'Come up to me on the mountain and stay here, and I will give you the tablets of stone with the law and commandments I have written for their instruction.'
Exodus 24:12

I have a friend, a skilled and experienced engineer, who spent a period of his career volunteering in Cambodia as part of the effort to clear landmines from the country. My wife and I visited him during his time there. We met some of his colleagues and learned more about their work. It turns out that there's a key distinction between military and humanitarian de-mining. If a military unit needs to cross a minefield, they simply have to clear a path wide enough to allow personnel and vehicles to pass safely across the mined zone. Humanitarian de-mining is more complex and involves clearing the entire mined area. If local land which has previously been mined is to be reclaimed so that children can play safely in that region, it is not sufficient to clear specific routes. The whole area needs to be made safe.

The reality is that we live at a time in which many places have not yet been completely cleared of mines, and when other locations pose related risks. Even walking on public footpaths on the south coast of England, when passing military firing ranges, I've come across notices warning of the dangers of straying from the path. Such signs focus one's mind. Knowing that I may trigger an explosion by deviating from the prescribed route can be a real incentive for keeping to the footpath!

As signs indicate risks of potential explosions, God's law warns us of danger. God is not a spoilsport. He does not forbid what is good. Instead, as the One who created both us and the world, he carefully warns of territory within which we may find we hurt either ourselves, others, or both. Through the boundaries of His law, He indicates those places in which we can move safely, and in which we will find freedom and life.

Soon after God rescued His people, Israel, from slavery in Egypt, He gave them His law.[1] The law set out the boundaries within which to ensure continued freedom. Refraining from murder was identified as a key aspect of being able to live freely. Committing murder both destroyed the life of the victim and wrought destruction within the life of the perpetrator too. Following the law was the way to keep within the bounds of freedom and to avoid the territory of harm.

As Paul, who wrote much of the New Testament of the Bible, was to reflect many hundreds of years later, the law wasn't enough.[2] After rescue from Egypt, a new form of slavery took hold, the slavery which came from being unable to keep the law. The law revealed the inability of God's people to live within these boundaries of abundant life.[3] God's people would need to be rescued again.[4] Chapter Seven of this book will say more of how Jesus, through His death and resurrection, went on to bring about that subsequent rescue. The Bible prompts us to look ahead to a time when there won't be any more tears or pain[5], when we won't need boundaries separating off what is life-giving from what is not, because everything which has endured will be life-giving, everything will be good. By that point all the mines will have been cleared from the whole territory.

In the meantime, the value of the law is in identifying the boundaries within which we find life-giving freedom. We face the challenge of living within those boundaries, but with the understanding that, in our own strength, we are unable to do so. To quote Paul again, 'I do not understand what I do. For what I want to do I do not do, but what I hate I do'.[6] Any experience we may have of failing to keep resolutions affirms what Paul is saying.

Temptations to live differently to the way God has called us are plentiful. Even as Moses was receiving the law from God in the form of the Ten Commandments, his people were making a blasphemous idol in the form of a golden calf.[7]

As Israel repeatedly turned her back on God in favour of human-made gods of wood, metal and stone, so we in our society are surrounded by temptations, some of them of our own making.[8] Money, sex and power can be alluring. We may experience technology too as a temptation. Kevin Kelly describes his feelings for the internet in language which evokes a sense of idolatrous worship:

> I am no longer embarrassed to admit that I love the internet … I find myself indebted to the net for its provisions. It is a steadfast benefactor, always there. I caress it with my fidgety fingers; it yields to my desires, like a lover … Rarely does it fail to please, and more marvellous, it seems to be getting better every day. I want to remain submerged in its bottomless abundance. To stay. To be wrapped in its dreamy embrace … The net's daydreams have touched my own and stirred my heart …. why can't you love the web?[9]

Kevin Kelly may be teasing us and there is nothing wrong with enjoying God's good gifts. The internet does have enormous potential for good.[10] Yet, at face value at least, Kevin's words seem to represent a step beyond appreciation of God-given gifts to a wrongful priority which takes our worship away from God, the only One to whom it truly belongs. There is a difference between enjoying pizza and focusing our lives around pizza, between celebrating music, and allowing passion for music to override all other priorities, between enjoying creation and treating creation as if it were the Creator.

Our own human inability to live in the way we have been designed to live is why, whatever the quotes from Yuval Noah Harari suggested in the previous chapter, human ingenuity alone cannot lead to fullness of life. Too many sources of trouble are rooted within us.[11] We need help to follow God. We can't do it in our own strength. Yet as Jesus declared, He did not come to abolish the law.[12] So God enables us through the Holy Spirit. It's only through this divine empowering that we find freedom. What we could not manage because of our own inability to keep to God's law becomes possible through God.

The Holy Spirit empowers us to live within the boundaries of the law. We now have the opportunity to live within those boundaries, not primarily through obedience to the law itself, but through growing in the character of our life-giving God whose very nature defines those boundaries in the first place. God grows His character within us, character of light and life, which by its very nature gravitates to those life-giving spaces marked out by the law.

So, what is the law? What are these life-giving boundaries?

Back in 2005, A J Jacobs began a year-long experiment in what he called 'living biblically' and attempting 'to follow the Bible as literally

46

as possible'. He recounts his experience in a book which is both amusing but also profoundly moving at times.[13]

One of many Old Testament laws which Jacobs followed was the command to Israelite men to leave the edges of their beards unshaven.[14] Over the course of a year his beard grew to be full and thick. Passers-by shouted 'Yo, Gandalf!', and ZZ Top was mentioned to him at least three times a week.[15] Yet Jacobs described his facial hair as 'simply the most noticeable physical manifestation' of his 'spiritual journey'.[16] The dominant theme of Jacobs' book, however, was the difficulty involved in attempting to follow the biblical law as if it were made up of random, disconnected, arbitrary rules. It's not.

God's law reveals God's character of love. That's why, when Jesus was asked to name the greatest commandment, He pointed to the two commands, to love God and to love one's neighbour.[17] This is how the gospel-writer Matthew records that exchange:

> Hearing that Jesus had silenced the Sadducees, the Pharisees got together. One of them, an expert in the law, tested him with this question: 'Teacher, which is the greatest commandment in the Law?'
>
> Jesus replied: '"Love the Lord your God with all your heart and with all your soul and with all your mind." This is the first and greatest commandment. And the second is like it: "Love your neighbour as yourself." All the Law and the Prophets hang on these two commandments.'[18]

As I explored in my previous book, I see these two commandments as part of a rich gift with which to illuminate issues of technology.[19] Through them, Jesus offers us powerful tools for a discerning engagement with technology. Following these two commands in our technological context involves considering how loving God with all our being, and loving our neighbour as ourselves, shape our understanding of a particular technology, or a technological issue.

I see the first of these two commandments as taking us deeper into the subject of the previous chapter. Surely loving the Creator God with all our heart, mind, soul and strength, involves honouring and respecting His own creativity. It involves engaging with His creation with an appropriate sense of awe and wonder. If I declare that I love the Creator God with all my being, yet I look upon human beings as second rate designs in need of upgrading, it seems to me that I seek to put myself in a place from which to judge the goodness of God's handiwork. God Himself has declared His creation to be good. Who am I, if I say that I love Him with all my being, to declare it to be anything else?

It's true that I can love a three-year-old child without believing that their artwork is worthy of being displayed in a national gallery. Yet to love God is inseparable from acknowledging His identity as the Creator of the universe. For a human to love God is for a creature to love their Creator. Such love precludes sitting in judgement on God's own creativity.

The second of these two commandments, to love our neighbour as ourselves, challenges us so that whatever our dreams of the future, we need to be thinking not just of ourselves but of others. We need to consider not just our own family, our own country, or our own group but all who are vulnerable and in need. This command prompts us to keep alert to the needs of any who may experience a sense of being an 'outsider' in one way or another, and of any who are in particular need.

48

These two commandments, to love God with all our being and to love our neighbour as ourselves, give us further biblical foundations with which to engage with our technological world. Such engagement is the subject of our next chapter.

SUMMARY

1. God's law provides protective boundaries for life in abundance.

2. We all experience temptations to live in ways which are not consistent with God's law.

3. It is only through the empowering of the Holy Spirit that we are equipped to keep God's law.

4. Jesus summarised God's law into the following two commands, which I will here call Biblical Truths Five and Six:

   a) Biblical Truth Five: We are commanded to love God with all our heart, mind, soul and strength.

   b) Biblical Truth Six: We are commanded to love our neighbour as ourselves.

5. Loving God as our Creator God involves honouring and respecting His creativity.

6. Loving our neighbour as ourselves involves looking outwards to the needs of others, not least any who experience being an 'outsider' or in particular need.

QUESTIONS FOR REFLECTION

1. How do you respond to the assertion of this chapter that God's law warns of potential hurt to ourselves or to others?

2. Do you think this assertion is also true of laws which relate to our relationship with Him, and to our worship of Him? Why?

3. Which idols or false gods seem to command most worship in our society today?

4. How easy do you find it to hold firm to any resolution which you make?

5. Is there a question you would like to ask A J Jacobs after his year long experience of seeking to 'live biblically'?

6. What links might we spot between the two great commandments identified by Jesus and the Ten Commandments recorded in Exodus 20:1-17?

1   The book of Exodus, the second book of the Bible, describes God's rescue of His people Israel from slavery in Egypt and His giving of the law.

2   In the biblical book of Romans, Paul writes to the early church in Rome and explores this theme, particularly within chapter 7.

3   The Old Testament witnesses to the inability of God's people to live in obedience to His law. Despite God's work through Priests, Kings, Prophets, Prostitutes, Servants, Foreigners, Insiders and Outsiders of all varieties, God's people failed to live in the way they were called by God to live. As well as describing this reality, the Old Testament also points ahead to the Messiah, the One who will come to rescue His people and to bring true and ultimate freedom.

4   The New Testament describes how, through the death and resurrection of Jesus, the Messiah, comes the ultimate rescue, of which the rescue of Israel from Egypt was a foretaste.

5   Revelation 21:4: '"He will wipe every tear from their eyes. There will be no more death" or mourning or crying or pain, for the old order of things has passed away.'

6   Romans 7:15.

7   Exodus 32.

8   A number of contemporary writers explore the idols and false gods of our Twenty-First Century society as a means of helping us recognise our own distorted priorities. In *God and the Pandemic*, p72, Tom Wright has written of how struggles to care for lives and the economy during the coronavirus pandemic might have been imagined as a clash between Asclepius, the Greek god of healing, and Mammon, the god of money. He notes that Mars, the Roman god of war, and Aphrodite, the Greek goddess of erotic love seem never far away. Archbishop Justin Welby has explored related themes within his Lent Book, *Dethroning Mammon*.

9   Kevin Kelly, *What Technology Wants*, p322-323.

10 Bill McKibben writes passionately of the internet as a resource for good, particularly in relation to climate change, in his book, *Eaarth*, p205-206.

11 G K Chesterton expressed this reality well, when in reply to the question 'What is wrong with the world?' he is reported to have responded 'I am.' https://www.chesterton.org/wrong-with-world/ Accessed 9th Sept. 2021.

12 Matthew 5:17: 'Do not think that I have come to abolish the Law or the Prophets; I have not come to abolish them but to fulfill them.'

13 A. J. Jacobs, *The Year of Living Biblically*.

14 Leviticus 19:27: 'Do not cut the hair at the sides of your head or clip off the edges of your beard.'

15 A. J. Jacobs, *The Year of Living Biblically*, p3.

16 Ibid., p3.

17 In naming these two Old Testament commandments, Jesus was quoting from Deuteronomy 6:5 'Love the LORD your God with all your heart and with all your soul and with all your strength', and Leviticus 19:18 'Do not seek revenge or bear a grudge against anyone among your people, but love your neighbour as yourself. I am the LORD.'

18 Matthew 22:34-40; Mark 12:28-34.

19 Justin Tomkins, *Better People or Enhanced Humans?*, p61-70.

4

A LIFE OF FREEDOM

The end of law is not to abolish or restrain, but to preserve and enlarge freedom.
John Locke[1]

As is the pattern within the even chapters of this book, it's now time to immerse ourselves in stories of our technological context once more in order to reflect upon that context in the light of the biblical truths of the previous chapter. Those were:

5. We are commanded to love God with all our heart, mind, soul and strength.

6. We are commanded to love our neighbour as ourselves.

Our contemporary technological context has roots which go back into times now past, and the stories of this chapter begin in Asia six hundred years ago.

***

In the 15th century the Chinese had built as impressive a fleet of ships as that found anywhere on the planet.[2] In 1414 the Chinese commander, Zheng He, set sail with a fleet 250 times larger than that with which Columbus would voyage to America almost eighty years later. Zheng He took 62 Galleons, any one of which could have contained all three of Columbus' ships. He travelled to Bengal, and brought home giraffes. He also visited Africa and Sri Lanka. The Chinese Navy was 'many times larger and more powerful than the combined maritime strength of all of Europe'.[3] Yet just a few years later the Chinese government took a conscious decision to scrap its navy and to pull out of all overseas trade and exploration. Zheng He's ships were soon decaying in their docks.

Why? China had internal concerns. It felt it had little to learn from the barbarian world outside, and it could do without foreign trade. That decision on an approach to technology may have led to the West rather than the East going on to dominate world trade over the next five hundred years. On the other hand, the Ming dynasty who ruled at that time and until the early twentieth century is said to have marked 'one of the great eras of orderly government and social stability in human history'.[4] Because of a technological decision, whilst much of Europe's energy went into exerting her influence around the globe, vast millions of Chinese people lived out their lives within a country focused upon self-sufficient national harmony.

One of the core questions raised by rapid technological change is whether or not it is possible for humans to slow down or even to halt such developments. Can we say 'Enough already!' to technology? The example above, of the Chinese giving up the technology of ship building, at least for a time, is one used by Bill McKibben in exploring this question.[5] He describes various examples of particular cultures saying 'Enough!' to specific technologies for a limited period of time.

Another fascinating example which McKibben gives comes from Japan.[6] In 1543 Europeans arrived in Japan, bringing, amongst other things, firearms. Within a decade, not only had samurai proved to be adept marksmen but Japanese craftsmen had learned to make firearms in large quantities and had improved upon certain aspects of the European designs. By the late sixteenth century guns were almost certainly more common in Japan than in any other country of the world. But then, in the early 1600s Japan did away with firearms. Why? Samurai started to recognise that guns were threatening the system of honour and civility which marked even the most violent samurai warfare. I don't want to be involved with either but I can imagine the distinction. Guns permitted anonymous killing at a distance not possible within the face-to-face encounters of samurai combat. And so, Japan did without guns for a period of three hundred years until the early twentieth century; three hundred years of dynamic cultural and economic growth.

I was amused to read Bill McKibben name the Amish as being 'the most technologically sophisticated people in North America',[7] but as I reflected on the statement, I began to sense what he might mean. The Amish have been prepared to say 'Enough' to technology in such a way that rather than embrace new technologies unquestioningly and individually, they assess different new technologies as a community and make a conscious shared decision about which technologies they will adopt and which they won't. Whether or not we agree with the boundaries they've drawn, I can see that they do indeed model a process of being discerning about the use of technology.

Maybe most significantly of all, Bill McKibben points to our world's use of nuclear weapons as another example of human ability to say

'Enough!' to technology.[8] Whilst we can never be complacent in relation to their use, it is indeed true that after bombs were dropped on Hiroshima and Nagasaki in 1945, we haven't seen nuclear weapons fired in conflict since.

***

In May 2016 the Oxford student newspaper, *Cherwell*, published a survey that showed that over 15% of students knowingly took *Modafinil* or another such drug without prescription whilst studying at Oxford.[9]

Modafinil, like *Adderall* and *Ritalin*, is an example of a 'smart drug', a pharmaceutical which may be used in order to seek to boost brain performance. Some of these drugs are prescribed treatments for narcolepsy, a condition causing the sufferer to fall asleep suddenly and at inappropriate times. They stimulate the central nervous system helping to keep the user awake during the day. Other 'smart drugs' may be prescribed as a treatment for attention deficit hyperactivity disorder (ADHD) and can help improve the concentration of children who struggle to pay attention at school. They are available on prescription for these purposes. They are put to a different use when used, without prescription, as 'enhancements' to enable highly-able university students to cram for exams.

Of the Oxford University students who declared having used a 'smart drug', only 12% said they had done so before arriving at the university. Oxford University is not alone. The use of 'smart drugs' by university students is said to be rife both within the UK and beyond. As the *Observer* newspaper put it, 'students used to take drugs to get high. Now they take them to get higher grades'.[10] The newspaper quoted Phoebe, a history student, as saying:

> It's not that it makes you more intelligent. It's just that it helps you work. You can study for longer. You don't get distracted. You're actually happy to go to the library and you don't even want to stop for lunch. And then it's like 7pm, and you're still, 'Actually, you know what? I could do another hour.'

***

The implantation of microchips inside the body is another example of an enhancement technology. In the summer of 2018, it was widely reported that several thousand Swedes had chosen to have microchips inserted into their hands in order to function as credit cards, key cards or even rail cards.[11] Some of those reports raised the question of what might make Swedes particularly likely to embrace such technologies. The reality is that the technology of microchip implantation is, of course, not restricted to those from Sweden. Sasha Twining, a reporter for *BBC South*, described her own experience of having a microchip inserted[12] and it was an Englishman who was the first ever to do so, back in the late 1990s. This is how that earlier event was reported:

> On Monday 24th August 1998, at 4:00pm, Professor Kevin Warwick underwent an operation to surgically implant a silicon chip transponder in his forearm. Dr. George Boulous carried out the operation at Tilehurst Surgery, using local anaesthetic only. This experiment allowed a computer to monitor Kevin Warwick as he moved through halls and offices of the Department of Cybernetics at the University of Reading, using a unique identifying signal emitted by the implanted chip. He

could operate doors, lights, heaters and other computers without lifting a finger.[13]

Professor Warwick later went further. He persuaded his wife to have a chip implanted in her own arm, which meant that the couple could communicate physically from a distance. When one of them squeezed their own hand, they stimulated a response in the other; holding hands electronically![14]

If Kevin Warwick's experiments were cutting edge for humans, at a similar time, around twenty years ago now, Belle was pushing the boundaries for her own species. Belle has been described as a 'telekinetic monkey'.[15] The Owl Monkey at Duke University in Durham, North Carolina had a chip implanted into her brain meaning that she could control a computer joystick using her thoughts. The work was funded by the *Defense Advanced Research Projects Agency* (*DARPA*), a military research organisation in the US. One can imagine potential military applications for fighter pilots or drone operators. Interestingly, Michael Goldblatt the scientist overseeing the work was motivated by the possibilities of the project leading to new treatments for children such as his daughter Gina who has cerebral palsy as a result of what she refers to as 'medical malpractice'.[16] Once more we come across a single developing technology with the potential for both therapy and enhancement.

More recently, on 28th August 2020, Elon Musk introduced the world to Gertrude the pig.[17] Surrounded by cameras and onlookers, Gertrude demonstrated the effects of having 1,024 electrodes monitoring the snout-controlling nerves in her brain. When those nerves fired, the electrodes detected the stimulation and triggered some music to play. When the observers around the world heard the musical jingle, they knew that Gertrude had found some food.

Gertrude's unveiling was a marketing performance for Musk's company, *Neuralink*, which applied in 2019 to launch human trials of related technology. Elon Musk argued that such implants could help cure dementia, Parkinson's disease and spinal cord injuries.[18] Others have questioned the privacy, political and commercial implications of this sort of neurotechnology.[19]

***

These stories raise a number of issues which aren't easily resolved on the basis of the biblical tools of the first chapter alone. For example, what limits might I place on my own engagement with technology? Is there a fundamental difference between a child with ADHD using *Ritalin* on prescription and a university student using it to cram for final exams? How might the insertion of microchips into my body, even within my brain, affect my understanding of my own humanity? The biblical command to love God with all our heart, mind, soul and strength may help us as we wrestle with these questions.

***

The previous chapter described how loving God might be understood as respecting His creative work, humans included. Limits and boundaries are part of that creation, not just the boundaries of the law, but also the inherent limits of the human form. The human form is limited in space by skin which marks the boundary of our body. Our existence, at least before death, is limited in time. We are born and at some later point we die. The physical capabilities of our bodies are limited. Without a device such as a plane or glider, our bodies themselves are not capable of flight. Even as athletes push the limits of human achievement year by year, there are at least conceptual limits to the speed at which we can run, the distance we can kick a ball, and so on. In a world and a body of limits, it seems appropriate that our technological engagement too involves the discerning search for appropriate limits. Whilst it may be beyond humanity as a whole to say 'Enough!' to technology, loving God with our heart, mind, soul and strength may inspire us to search for wise limits in our own personal technological engagement.

* * *

Loving God with all my heart, mind, soul and strength may also affect my response to innate human limits. The issue of 'smart drugs' brings us back to the question we first touched upon within Chapter Two of a possible distinction between therapy and enhancement. If I understand loving God my Creator to involve respect for the way He has created humans and the world, such a distinction may be important to me. I may wish to distinguish between innate limits which God has chosen to build into His good creation, and particular and personal limits which may simply result from the consequences of the fact that creation is marred.

For example, if I have ADHD, I may look around at others and see that such a condition is not an inherent aspect of being human. It's widely understood to be a disorder which can be treated with a drug such as *Ritalin*. Taking such medication is then a therapeutic solution to a disorder which requires healing. If on the other hand, I am a university student seeking to use *Ritalin* to improve my already high levels of concentration, I am pursuing enhancement. Such enhancement might be challenged on the basis of any dissatisfaction it may express for innate human limits which God has built into His creation. Therefore, whilst valuing the healing of marred creation, loving God and respecting the goodness of His creativity might prompt us to resist using a healing drug for an enhancement purpose.

I recognise that the topic of 'Human Enhancement' provokes more questions than can be dealt with simply by distinguishing between therapy and enhancement. The subject continues to provoke rich ethical engagement.[20] Nonetheless, I am convinced that discerning between therapy and enhancement is a critical aspect of reflecting upon the use of enhancement technologies. Whilst the use of healing technologies can at least be affirmed on the basis of the need for healing within marred creation, the use of enhancement technologies may be questioned on the basis of whether they express a desire to improve upon the human form. If they do, it may be helpful to reflect upon whether such uses are consistent with loving God with heart mind, soul and strength.

***

We might reflect upon the use of microchip implants on exactly this basis. If, as Elon Musk proposes, such implantations were used to treat conditions such as dementia or Parkinson's disease, we might at least affirm them as a healing act. If instead, these implants were used to seek to improve upon the human form which God has created, as if it were a flawed model in need of an upgrade, we might choose to challenge such use for that reason.

***

What about the second biblical command of the previous chapter, that of loving our neighbour as ourselves? Let's immerse ourselves in some more stories before we consider how they might be illuminated by that second command.

Paro is a robotic seal developed in Japan as a companion for the elderly.[21] It has soft fur and has been programmed to nurture a bond of companionship between seal and owner. That programming appears to have been effective. Trials in numerous homes for the elderly have found that robots such as these provide a meaningful source of companionship for those who spend time with them. *Paro* is not the only such robot available. *Aibo*[22] and *MiRo*[23], both robotic dogs, have also already been used extensively as companions. Humanoid robots, such as the *Care-O-bot*[24] and *Pepper*[25], have been developed to provide more complex and comprehensive care.[26]

These devices raise the question of the value of human involvement in such care. Are these devices primarily about giving joy and meaning to the lonely? Or are they instead tools designed for enabling family and friends to delegate their own care to a machine, and so avoid feelings of guilt whilst staying absent themselves? A group of children learning about such machines asked the question 'Don't we have people for these jobs?'[27]

We may be tempted to question the idea of a fully grown adult experiencing a meaningful relationship with a robotic soft toy. Nonetheless, there is much evidence that whenever we encounter human-type behaviour in an object, or an animal, we have a natural tendency to respond as if it were human. *Eliza* was a computer programme written in 1964[28] to act as an electronic therapist. It worked through identifying words of emotion and asking more about those emotions, or simply replying with open comments such as 'Please go on'. Despite being completely aware that *Eliza* was a computer programme, users would ask to be left alone with 'her' in private for hours at a time, and later reported having had a meaningful therapeutic experience.

***

Susan Maushart, who disconnected from screens for six months as described in Chapter Two, playfully describes not just a meaningful engagement with technology but a fully formed relationship. This is how she writes about her feelings for her phone before the start of her family 'Experiment':

> We'd only been together for six months, but in that time we'd developed a relationship that was totally in sync, in a totally out-of-sync kind of way ... Talk about a toxic breakup. This one had all the elements: anger, denial, bargaining ... I loved Della – yes, my laptop had a name (shaaaame!) – but I also recognised that a trial separation was probably the best thing that could happen to us at this stage in our relationship.
>
> Looking back, I can see that Della was the spouse and helpmeet – faithful, reliable, comfortable, and just a little dull. But my *iPhone*, iNez? Hoooo, mama! Now that was one smokin' hot affair. In one impossibly sexy handful, iNez embodied all the things I love best about technology. It doesn't get any less humiliating when I stop to think about exactly what those things are. Basically, iNez was compliant, discreet, entertaining, ridiculously receptive, and looked amazing in black lacquer. She might as well have been a freaking geisha. If I'm honest – and it's killing me to admit this to myself, let alone to you – I got a buzz from being seen with iNez. I loved what she could do, but I also loved what she stood for – some heady confluence of youth and wealth and mastery. Being seen in her company made me feel important, powerful, "in the loop." But loops have a way of tightening gradually. That's why I had to leave her.[29]

She continued:

In the early years, once my workday was over … so was my email attention span. It never really occurred to me to go racing back to Outlook after dinner or before bedtime. … Once I hooked up with iNez, those relatively functional patterns dissolved. Now that I had the freedom to walk around in-boxicated (*sic*) all day long (and all night too if I wanted), my info-neediness went through the roof.[30]

Even as we smile at the expression of these lines, how many of us can say that we can't empathise with aspects of the experience behind these words? A number of science fiction writers take these ideas and extrapolate them into the future. Both *Bicentennial Man* starring Robin Williams, and more recently *Her,* with Joachin Phoenix explore the possibility of a person falling in love with a machine. These and other films raise questions about how we treat machines, particularly if they appear to have human-like qualities. Will machines ever develop feelings of their own? How would we respond if they did so? Would we know the difference between a machine which had feelings and one which simply behaved as if it did?

***

Love affairs with phones may be a relatively recent phenomenon but passion for cars is nothing new. Self-driving car technology has come a long way in the last few years. Parking sensors and cruise control are now commonplace within new automobiles. Automatic parking is not unusual. Self-driving vehicles have clocked up millions of test miles. There are even plans in place to race self-driving cars around the famous Indianapolis racetrack in 2021.[31]

As of October 2018, *Waymo*, the company that grew out of *Google*'s self-driving car project, announced it had completed 10 million test miles.[32] At that point *Uber* had completed 2 million miles[33] though its progress had been put on hold after its involvement in a fatal collision in March 2018.[34] That crash took place in Arizona and involved a *Volvo*, operated by *Uber*, hitting a pedestrian, Elaine Herzberg, as she wheeled a bicycle across the road. The self-driving *Volvo* correctly identified the object as a bicycle and recognised the need to put on the emergency brakes but the machine was not set up to apply the brakes autonomously. *Uber* employed people to sit in the cars and to monitor both the vehicle's alerts and the surrounding area. It was down to those human monitors to physically apply the brakes when alerted to do so by the vehicle. In this case, the person failed to respond. She is thought to have been streaming a television programme and so not to have seen the warning in time to act.

It's easy to recognise that if a human monitor is to be effective in that sort of situation, they need to be paying attention. It's not straight-forward though for a person to maintain such focus and it raises the issue of how often their intervention might be necessary. If a human has to take control every five minutes, they'd find it easier to stay focused than if they're only needed every five hours, or every five days. What if the vehicle is developed to such a high level of autonomy that it only involves a human interaction every five months, or five years? The less often that human intervention is necessary, the less likely it is that a human will be alert to respond in that instance. Whether to split authority between humans and machines on the road, and if so, how to accomplish that safely, is a challenging issue for both manufacturers and legislators.

The issue of whose safety to prioritise presents a further complex challenge. Do we want cars which will protect human life as far as possible, no matter where that life is, or do we want cars that offer preferential protection to those inside the car? When car manufacturers put that question to the general public, they get a clear response saying that as humans we want all human life to be protected, whether that's the pedestrian crossing the road, or the passenger in the car.[35] Prioritise human life. But when the question is changed to what sort of car would you buy, the one which aims to protect human life, whether inside or outside the car, or the one which prioritises the life of those inside the car over those outside, then car manufacturers receive a different response.[36] We want the world to be one way, but we'll spend our money influencing it another way, in order to benefit ourselves and our family.[37]

We may not yet be at a place to implement such priorities, and the realities of programming will mean that the internal processes of a self-driving car will of course be very different to that which might go on within a human brain. Nonetheless, these discussions do illustrate how technology and ethics go hand in hand. Who benefits from technology? Whose decisions influence that outcome?

**

The biblical command to love our neighbour as ourselves illuminates each of the previous stories.

Robots such as *Paro* might either be affirmed or challenged on the basis of the command to love our neighbour.[38] Understood as devices to provide care which would otherwise be unavailable, they might be affirmed on that basis. If instead, they are used to abdicate personal responsibility for offering care they might be challenged.

The children who asked 'Don't we have people for these jobs?' highlight the central issue of how robotic support relates to human care. In what circumstances will we choose to provide care for the elderly robotically instead of directly through a human being? Is robotic care offered over and above direct human care, without any diminishment of that human care? Does the robotic care change the nature of the human involvement from direct care to administrative oversight of the machine? Is the robotic care primarily meeting the needs of the elderly person themselves or of their family? How do economic structures in our society affect a person feeling pulled between the demands of paid employment and caring for a friend or relative in need? What's the impact of the robotic involvement upon the person being cared for and upon their relationships with family and friends?

The command to love our neighbour as ourselves might illuminate reflections on these questions and help ensure that any machine care which is deployed is offered as an act of love rather than in avoidance of neighbourly care.

***

Ever greater immersion in a machine world may threaten our connection with one another as neighbours. We may recognise the roots of such a scenario in our world today.

A similar outcome in terms of human disconnection might come about in other ways. *Elysium*, the 2013 film starring Matt Damon and Jodie Foster, portrays a world in which rich and poor are ruthlessly separated. The poor remain on the barren wasteland of what once was Earth, whilst the rich live in a luxurious space station. The physical gap between rich and poor has been deliberately enlarged to the point where it is almost impossible for either to be a good neighbour to the other.

Any future in which our ability to love our neighbour is restricted, whether that's by behaviour, structures or anything else, might be questioned on that basis.

***

Susan Maushart's connection with her mobile phone may have had a powerful influence upon her but she describes it as a relationship for comedy and effect, not because she truly believes that in that technology she found a soul-mate. The relationships portrayed in the films *Her* and *Bicentennial Man* are very different. In both cases, the movies explore the lives of people who truly do believe that in a software programme or a robot they have found the love of their life. How might we reflect upon these scenarios and is it even worth doing so?

We'd be rash to dismiss the idea of such attraction. As we see faces in clouds, we do project feelings onto machines, particularly humanoid ones. It seems to me that science fiction films and television shows provide one of the key ways in which our society reflects upon such issues of our technological future. Day to day news on technological developments comes so fast that it can be hard for us to keep up to date. We may be aware of our lack of understanding of key issues involved. We may have a sense of legal, political, scientific and commercial experts working in the background to ensure that developments are guided and controlled. We may have a sense of powerlessness in affecting change ourselves. And each little development is so incremental that it may not seem worth bothering about anyway. Yet movies of technological futures give us the freedom to step back, to laugh at what is ridiculous and to pause to reflect upon where we might be headed. They may reassure us that 'at least we aren't there', or caution us if a dystopian future feels worryingly familiar.

I listened to a podcast recently in which the writer and speaker, Kevin Roose, was interviewed about his book *Futureproof*.[39] Kevin commented on how he was more confident in the potential benefits of technologies such as Artificial Intelligence than in those who might direct the use of those technologies. He prompted me to reflect once again on the significance of our role as humans in shaping our technological context. Some of us will do that as engineers, health workers and entrepreneurs. Others will contribute through a legal or political framework. Others of us will do so as consumers, parents, voters and through prayer. My sense is that reflecting upon the future, including through the creativity of fiction, can help us to use well whatever influence we may have in shaping the coming years.

So how does the command to love our neighbour speak into the scenario of a man who thinks he's in love with the operating system within his phone? Primarily, as we noted earlier, we might challenge any society which allows human-machine interactions to take the place of relationships between humans. This seems to me to be more than enough reason to caution against forming a romantic relationship with a machine, or seeking one as a spouse. If we live in such a way as to disconnect ourselves from human contact that might be challenged on the basis of making it harder to obey the command to love our neighbour as ourselves.

All this applies regardless of any question of a machine's sentience or lack of it. People disagree on questions regarding the extent to which various animals are sentient. How much more complicated might discussions become in relation to machines? Yet, we don't need to be able to answer such questions in order to challenge any relationships with machines which make it less likely that I am obedient to the command to love my human neighbour as myself.

Are there ways though in which we might appropriately experience a machine as a neighbour? When Jesus was asked the question 'Who is my neighbour?', He told the story of the Good Samaritan and then asked who behaved as a neighbour.[40] Jesus flipped the question around from who do I recognise as a neighbour to what must I do to be a neighbour to others. Bioethicists considering the human embryo have used that thinking to flip the question from 'Is this embryo my neighbour?' to 'What would it mean for me to be a neighbour to this embryo?'[41] Might similar considerations help us with machines?

There seems to me to be an immediate implication for any violent or abusive behaviour towards machines, as explored in much science fiction. If a machine looks like a human being, I don't need to know whether or not that machine is my neighbour in order to know what it means for me to behave as a neighbour towards it. The way I treat a robot which looks like a human being will have an effect upon me, regardless of the machine's sentience or lack of it. If my 'mistreatment' of a machine conditions me to be more likely to behave violently or abusively to a human person, then such behaviour towards a machine might be challenged on the basis of the command to love one's neighbour. There might be situations, of course, where such a challenge is overridden by a greater concern. For example, a faulty humanoid robot might need to be forcibly prevented from causing harm to others.

***

Questions of how to design a self-driving car which behaves ethically, who decides what those ethics are, and how might such a system be policed all raise challenging issues. Nonetheless, the command to love one's neighbour as oneself draws attention to the question of who benefits from technology. Seeking to be obedient to that biblical command will continually prompt us to consider the impact of our own technological engagement upon others.

***

God's law is a gift which enables us to be discerning in our engagement with technology. As Israel discovered, however, God's law alone is not enough. God sent His Son, the Messiah, to set His people free. In the next chapter we will explore how God being born as a human baby affects our thinking about what it means to be human in our technological context.

SUMMARY

1. There are examples of particular peoples saying 'Enough already!' to particular technologies for particular times (*i.e.,* in fifteenth century China, seventeenth century Japan and in modern day Amish communities in North America).

2. 'Smart drugs', available on prescription to treat conditions such as narcolepsy, may be used by others to seek to enhance brain performance.

3. Implanting microchips into humans and animals has been used to store data, manipulate electrical devices and even to affect brain function.

4. The command to love God with all our heart, mind, soul and strength may be used to reflect upon technology in relation to innate limits within ourselves and our world.

5. *Paro*, a robotic seal, and similar devices are being used as robotic companions for, amongst others, the elderly.

6. Susan Maushart has written humorously of developing a 'relationship' with her mobile phone. The films *Bicentennial Man* and *Her* extend such a concept to the realms of humans falling in love with machines.

7. The programming of self-driving cars raises ethical issues about protecting human life both inside and outside the car.

8. The command to love one's neighbour as oneself may challenge us to ensure that we address the needs of others as well as of ourselves within our technological discernment.

QUESTIONS FOR REFLECTION

1. Do you feel it is possible for humanity as a whole to say 'Enough already!' to technology? Why?

2. If they were freely and legally available, can you think of situations when you would choose to use 'smart drugs'?

3. Does the idea of an embedded microchip to store data within your hand appeal to you? Why?

4. In what ways have you experienced machines moving you closer to or further away from other people?

5. Have you experienced yourself investing deep emotions in a 'relationship' with a machine?

6. What might be the pros and cons of treating a machine as a person?

7. Would you consider buying a self-driving car which might sacrifice the life of the driver and its passengers for the sake of those outside the car?

1   John Locke, *The Second Treatise of Government*, Chapter 6, Section 57: https://www.gutenberg.org/files/7370/7370-h/7370-h.htm Accessed 9th Sept. 2021.

2   Bill McKibben, *Enough*, p174-176.

3   Ibid., p 174.

4   Ibid., p176 – McKibben attributes this quote to John K Fairbank.

5   Ibid.

6   Ibid., p176-179.

7   Ibid., p171.

8   Ibid., p179.

9   https://cherwell.org/2016/05/13/revealed-oxfords-addiction-to-study-drugs/ Accessed 9th Sept. 2021.

10 https://www.theguardian.com/society/2015/feb/15/students-smart-drugs-higher-grades-adderall-modafinil Accessed 9th Sept. 2021.

11 https://www.independent.co.uk/voices/sweden-microchips-artificial-intelligence-contactless-credit-cards-citizen-science-biology-a8409676.html Accessed 9th Sept. 2021.

12 https://www.bbc.co.uk/sounds/play/p071fs48 Accessed 9th Sept. 2021.

13 http://www.kevinwarwick.com/project-cyborg-1-0/ Accessed 9th Sept. 2021.

14 https://www.atlasobscura.com/articles/nervous-system-hookup-leads-to-telepathic-hand-holding Accessed 9th Sept. 2021.

15 Joel Garreau, *Radical Evolution*, p19.

16 Ibid., p17.

17 Laura Sanders, *Inside your Head*, p24-28, *Science News*, February 13, 2021.

18 https://www.bbc.co.uk/news/world-us-canada-53956683 Accessed 9th Sept. 2021.

19 Laura Sanders, *Inside your Head*, p24-28, *Science News*, February 13, 2021.

20 One small contribution to that debate is my 2013 book, *Better People or Enhanced Humans?*

21 http://www.parorobots.com/ Accessed 9th Sept. 2021.

22 https://us.aibo.com/ Accessed 9th Sept. 2021.

23 http://consequentialrobotics.com/miro-beta Accessed 9th Sept. 2021.

24 https://www.care-o-bot.de/en/care-o-bot-4.html Accessed 9th Sept. 2021.

25 https://www.softbankrobotics.com/emea/en/pepper Accessed 9th Sept. 2021.

26 https://www.independent.co.uk/life-style/gadgets-and-tech/features/robot-carer-elderly-people-loneliness-ageing-population-care-homes-a8659801.html Accessed 9th Sept. 2021.

27 Sherry Turkle, *Alone Together*, p76.

28 Brian Christian, *The Most Human Human*, p75.

29 Susan Maushart, *The Winter of our Disconnect*, p102-103.

30 Ibid., p107-108.

31 https://www.wsj.com/articles/autonomous-vehicles-to-race-at-indianapolis-motor-speedway-11595237401 Accessed 9th Sept. 2021.

32 https://techcrunch.com/2018/10/10/waymos-self-driving-cars-hit-10-million-miles/ Accessed 9th Sept. 2021.

33 https://www.forbes.com/sites/bizcarson/2017/12/22/ubers-self-driving-cars-2-million-miles/#786438eca4fe Accessed 9th Sept. 2021.

34 https://www.nytimes.com/2018/03/19/technology/uber-driverless-fatality.html Accessed 9th Sept. 2021.

35 Hannah Fry, *Hello World*, p126. Fry quotes a survey published in the journal *Science* in 2016 which found that '76% respondents felt it would be more moral for driverless cars to save as many lives as possible.'

36 Ibid., p126. Fry quotes the same survey by *Science* finding that when respondents were asked if 'they would actually buy a car which would murder them if the circumstances arose, they suddenly seemed reluctant to sacrifice themselves for the greater good.'

37 https://spectrum.ieee.org/cars-that-think/transportation/self-driving/people-want-driverless-cars-with-utilitarian-ethics-unless-theyre-a-passenger Accessed 9th Sept. 2021.

38 I am assuming that the electronic therapist *Eliza* was never intended to replace the therapeutic care offered by a human professional but was instead designed to push back the developmental boundaries of such technology. Whenever such devices are used in place of a human therapist then the discussion around *Paro* applies here too.

39 Kevin Roose, *Futureproof*.

40 Luke 10:25-37.

41 Oliver O'Donovan drew upon the parable of the Good Samaritan in his 1984 book, *Begotten or made?*, p49-66. He argued that discerning 'Who is a person?' involves expressing our own personhood. Some years later Ian McFarland proposed a similar focus upon what it means for us to behave as a person to others. See his chapter within the 2002 book, *Theological Issues in Bioethics* edited by Neil Messer, p76-84.

5

THE INCARNATION

'You will conceive and give birth to a son, and you are to call him Jesus'.
Luke 1:31

I'm a fan of *Formula One*. I like the noise and the spectacle of the cars. I enjoy Lewis Hamilton's success. I am intrigued by the way the contributed talents of engineers, mechanics and other team personnel are combined with the skills of a specific driver. I am inspired by the push for marginal gains in a striving for excellence. I delight in the statistics! I love having inherited a passion for the sport from my father and having been taken by him, as a toddler, to a number of races.

I had the privilege of attending part of a Grand Prix race weekend with my own son a few years ago. One of the things that struck me was the sheer number of people wearing team caps or driver's T-shirts, or carrying logo-laden umbrellas or bags. As with any sport, the association of oneself with the focus of one's support is part of the joy and the pain of the fan experience. For many, whether in a football stadium, cricket ground, or at a motorsport track, wearing the appropriate merchandising is a key aspect of that association.

For popular entertainment, contemporary films depict Norse gods such as Thor entering fully-formed into the world of humans.[1] When

God Himself entered into His own creation He chose a different way. At the heart of the good news conveyed by the Bible lies the truth that God Almighty, the Creator of the universe, became human out of love for us.[2] He clothed Himself with the stuff of His own creation. He embodied Himself with human flesh. What's more, he did so, like us, not immediately as a fully grown human adult, but through being born as a vulnerable human baby.

The Bible describes how a young woman, Mary, was visited by an angel who explained to her that, through the action of the Holy Spirit, she would conceive and give birth to a child, Jesus.[3] Former Archbishop of Canterbury, Rowan Williams, has written of a 'microscopic change in the physiology of a young woman in Nazareth that is the hinge on which the history of the world turns'.[4] When the Creator of the universe entered into His creation He did so, growing and developing within the womb of Mary, before being born and placed within a manger.

As is celebrated around the world each Christmas, Jesus was born with all the fragility of a new-born human baby. Not only did Jesus enter into this world with that physical vulnerability but He did so in an occupied land, to poor parents, at risk from a murderous dictator.[5] Yes, Jesus 'made himself nothing by … being made in human likeness',[6] yet in doing so He showed God's strength in human weakness.[7] He revealed His sovereignty and His control over all that threatens to harm or to destroy.

Furthermore, Jesus grew at the pace of childhood development, demonstrating patience and trust in God's timing. The relative quiet of the biblical record in relation to Jesus' growth and development[8] testifies to that patience and to the goodness of the rhythms and seasons of life and of the world.

Another key aspect of Jesus' humanity is His relationship with others. Jesus was raised in a family with His mother Mary, her husband Joseph, and with brothers and sisters. He was known by the local community. Later on, during His public ministry he journeyed with others, notably the twelve disciples. The Bible tells us of occasions when He spent time with a smaller group of three: Peter, James and John. We also read of a larger group of men and women who travelled with Him and cared for Him. Jesus' earthly life was a life lived in community.

This fact of the Incarnation, as it's called, illuminates every aspect of our lives, for it means that whatever situation we find ourselves in, we are not alone. Jesus, Sovereign Lord of the cosmos, understands what it is to be human. We don't worship a God who is absent or remote. Jesus knows hunger, tiredness, temptation and pain, as well as joy, friendship and laughter. He has demonstrated his solidarity with us. Jesus knows what it is to be human. He understands us. Jesus walked and worked and slept and wept. Our human lives already possessed immense value through having been made by God and declared not just good but very good, and through having been made in the image of God. Yet they became of even more worth, when Jesus, the Creator of the world, chose the human form through which to become part of His own creation. It is through Jesus becoming human that He rescues us from all that might separate us from Him.[9] Jesus roots for us and cheers us on, far more than even the most dedicated sports fan does for the focus of their support.

So, when we think about our technological context, and our engagement with that world, our physical embodiment is significant. When we reflect upon the fact that we are embodied creatures and when we think about the physical stuff of creation, we do so knowing that in Jesus, God Himself became embodied within Creation. Jesus demonstrated, in the most powerful way possible, that our physical world and our human bodies matter. They are created by Him, they are made in His image, and they are the form chosen by Him with which to enter into His own creation and to express His solidarity with us.

In Chapter Three we considered how the command to love God involves honouring and respecting His creation, humans included. Loving God precludes seeing our own embodiment as second rate, or in need of improvement. The fact that God has created us is already more than enough to cause us to respect and to honour the way in which He has done so. This conclusion is further reinforced through the Incarnation. The reality that God Himself became human and chose that form for His own embodiment adds yet more weight to the existing significance of our humanity.

As Jesus embraces human life, He calls us not just to live, but to live a life which is full to overflowing with His abundant goodness:

> I have come that they may have life, and have it to the full.[10]

Jesus calls us to be fully alive as a human being and He Himself models what it is to be so. We're not called to tolerate our humanity, or to grudgingly accept our human frailties, but with humility to delight in the extraordinary experience of what it is to be human.

***

The *Loebner Prize*[11] is a Turing Test[12], an eponymous test proposed by Alan Turing in 1950 as a means of assessing machine intelligence. This particular Turing test is designed to find the piece of software most able to convince human judges that the computer running the software is human. The test works by having a panel of human judges interact via a computer terminal with other humans and a number of machines. The task of the judges is simply to distinguish between the humans and the machines. Each year a prize is given to the computer which most successfully fools the human judges into believing that it is a human being. A prize is thus awarded for the 'Most Human Computer'.

The judges experience a profoundly enhanced version of a rare human condition called phonagnosia, or voice blindness. Someone suffering from the condition can tell from your voice if you're male or female, young or old, but they can't identify who you are. Steve Royster has the condition and even when his mother calls him on the phone, he can't recognise her from her voice.[13] This is not unlike the position within which all of us can find ourselves when it comes to the internet. We need to learn how to work out, is this email that's supposedly from my friend, saying that they're stuck in Thailand with no money, really from my friend or am I being scammed?

It's not always easy to avoid being caught out. Robert Epstein, an American psychologist from the University of California in San Diego, has written about his experience of falling in love with Ivana from Russia.[14] They met on a dating site. He acknowledges that he was first attracted to her picture. Her English wasn't good but over a period of months they corresponded by email and he fell in love. But then he started to note that things weren't quite right. She'd tell him about having walked in the park that day and yet when he looked up the weather for her home, he was able to discover there had been blizzard conditions. When he then sent a nonsense message she replied without comment. Not long afterwards he was able to confirm that not only was she not a woman, but she wasn't a person at all. She was a computer programme or 'chatbot'. As he later reflected on the experience, he commented that he had such a desire to be loved that he missed the cues which might have opened his eyes to the hoax earlier. We can only imagine whether or not, in his situation, we might have been fooled ourselves.

The journalist Brian Christian has reported on the history of the *Loebner Prize* and his experience of being one of the human participants within the 2009 competition.[15] In his 2011 book on the subject, Brian describes his exploration of how best to distinguish himself from the machine mimics. I enjoyed his discovery that early 'chatbots' couldn't easily cope with non-words. For example, programs which responded to the question 'What do you think of dlwkewolweo?' with anything other than befuddlement, might quickly be identified as machines.[16] Similarly, chatbots might also be unlikely to respond to other 'non-words' such as 'um' and 'ah' in a human-like way. It struck me how our humanity isn't all about our sophistication and our articulateness. The failings of those early chatbots invite us to embrace and to accept all that makes us fully alive, mumbling and all!

Yet, Brian Christian was also discerning in seeking to be identified as human through distinctly positive qualities. He describes how one of the humans who was most successful at distinguishing himself from the machines in an early competition did so through being 'moody, irritable and obnoxious.'[17] Brian Christian felt this to be not only bleak, but also a call to arms.

At the end of the competition in 2009, one of the software programs, written by David Levy, won the prize for being the 'Most Human Computer'.[18] Brian Christian won the prize for being the 'Most

Human Human', the one of the four human participants who was most easily identifiable by the judges as being human![19]

Brian's book is a fascinating account of how machines mimic humans and of how we might choose to respond. He suggests that whenever we experience machines growing in the ability to act like people, we face a challenge. We might relinquish that activity as one in which humans excel, and so yield that territory to the machines. We say that we've been beaten at chess and we'll allow that chess ceases to be a human activity. Or we take a different route, and we say that even as machines mimic human behaviour, we will rise to the challenge and we will work out what it means for us to become yet more fully human.[20]

Within discussions on science and faith, the term 'God of the gaps' has been used to warn of the consequence of perceiving evidence for God only within gaps of scientific knowledge.[21] If God's activity is seen as limited to those gaps, then as scientific knowledge grows, God's activity is seen to shrink. If instead, God's activity is understood throughout and beyond creation, as the Bible describes it, growing scientific knowledge is not a threat to, but rather an illumination of that work. In the same way, there are consequences of seeing ourselves as 'people of the gaps', defined by the limitations of machines. If we value only those human characteristics which are not yet successfully mimicked by machines, our identity will shrink as technology develops.

Brian challenges us to take a different route, and to celebrate our humanity, regardless of machine capability. Interestingly, even though in 2008 the human participants had lost five votes to the machines, in 2009 none of the machines fooled human judges into identifying the human participants as machines.[22] Even as the machines had improved, so too had the humans.

Brian Christian raises the question of what it means to be fully alive as a human being in our technological world. He may have won the 2009 *Loebner Prize* for the 'Most Human Human' yet it is Jesus who most truly embodies that title. When we seek to respond to Jesus' call to be fully alive as a human being, He provides us, in Himself, with a perfect example of just what that means. In Chapter Seven we will explore how Jesus went on to die and to rise again. Before then, we will consider how the reality of Jesus' humanity offers a way of shaping how we live our own lives in our contemporary context.

SUMMARY

1. Biblical Truth Seven: In Jesus, God Himself became a human being.

2. In this way, God further affirmed, and gave value to, our own humanity.

3. Because of this, God knows and understands all that it is to be human.

4. Biblical Truth Eight: Jesus grew at the pace of human childhood.

5. In this way, Jesus demonstrates patience and trust in God.

6. Biblical Truth Nine: Jesus' earthly life was a life lived in community.

7. When machines begin to mimic human behaviour, we face a choice, either to relinquish those activities to machines, or to become more fully human ourselves as we continue to engage in them.

8. Biblical Truth Ten: Jesus calls us to be fully alive as human beings and He models for us what it means to be so.

QUESTIONS FOR REFLECTION

1. If you support a sports team, or other group or individual, in what ways do you express that support?

2. Does it surprise you that God had the patience to grow as a human child? Why?

3. Do you find it hard to believe that someone else could understand every aspect of your life?

4. If you were typing away as part of a competition for the *Loebner Prize*, how might you try to convince a judge of your humanity?

5. How do you feel about the fact that it took Robert Epstein so long to recognise that *Ivana* was a chatbot?

6. Does your identity as a human feel threatened when machines achieve ever higher levels of human mimicry, through chess, speech, movement, and in other ways?

7. How can we model our lives on Jesus' life, whilst living the unique and particular life which He calls each one of us to live?

1   For example, various films within the *Marvel Cinematic Universe*.
2   John 1:1-14.
3   Luke 1:26-38.
4   Rowan Williams, *Candles in the Dark*, p1.
5   Matthew 2:1-18.
6   Philippians 2:7.
7   2 Corinthians 12:9: 'But he said to me, "My grace is sufficient for you, for my power is made perfect in weakness." Therefore I will boast all the more gladly about my weaknesses, so that Christ's power may rest on me.'
8   The four biblical gospels, Matthew, Mark, Luke and John, between them tell us only very little about Jesus' life before the age of around thirty when His public ministry began.
9   Thomas F. Torrance in his book *Incarnation* refers to the early church father, Gregory Nazianzen. In quoting the phrase, 'the unassumed is the unredeemed', he describes the rescue of our human bodies arising from God in Jesus entirely assuming, and so entirely redeeming, our humanity.
10 John 10:10b.
11 https://www.ocf.berkeley.edu/~arihuang/academic/research/loebner.html Accessed 9th Sept. 2021.
12 https://www.britannica.com/technology/Turing-test Accessed 9th Sept. 2021.
13 Brian Christian, *The Most Human Human*, p16.
14 Ibid., p9.
15 Ibid.
16 Ibid., p156-157.
17 Ibid., p5.
18 Ibid., p260.
19 Ibid., p260.
20 Ibid., p260-266.
21 See, for example, John Polkinghorne, *Science and Creation*, p13-15.
22 Brian Christian, *The Most Human Human*, p261.

6

A WELL-FITTING LIFE

It's not a faith in technology. It's a faith in people.
Steve Jobs[1]

The biblical truths from the previous chapter were:

---

7. In Jesus, God Himself became a human being.

8. Jesus grew at the pace of human childhood.

9. Jesus' earthly life was a life lived in community.

10. Jesus calls us to be fully alive as human beings and He models for us what it means to be so.

As we turn now to our contemporary context, we pick up the story of *IBM*'s *Watson* where we left it in Chapter Two, at the point of the machine's quiz show triumph.

***

The seven-year path to the television show-down had been successfully completed by the *IBM* team, but the journey didn't end there. It had become clear for some time that the tools which *Watson* had acquired to compete at *Jeopardy* were transferable.[2] Medicine was one field in which *IBM* had seen the potential for deploying *Watson*'s impressive talents. By 2013, developments of *Watson* were being used to assist doctors with diagnoses and treatment decisions.[3] It is noteworthy that *Watson* wasn't designed to replace a human, but rather to augment their ability, equipping the doctor to carry out their work even more effectively.

There is an interesting parallel between this fact and the experience of Garry Kasparov after being beaten by *Watson*'s predecessor *Deep Blue*. As Garry reflected back upon his own experience of machine-inflicted defeat within a field which, up until that time, had been a uniquely human domain, he noted how person-machine competition in chess had led not to the dominance of one over the other but to the emergence of a new cooperation.[4] 'Advanced Chess', also known as 'Centaur Chess', involves humans competing against one another, not simply drawing upon their own wits, but with real-time access to computing power in order to improve their game.

Watson's deployment in the field of medicine has not been straightforward. There's still much room for improvement with these technologies.[5] Nonetheless *Watson* is credited with diagnosing a woman in Japan with a rare form of leukaemia, and of discovering five genes linked to motor neuron disease.[6] In the same field, over the last few years, robotic surgery has been developing, another example of human-machine collaboration.[7]

The human-machine collaboration in which *Watson* participates, and seen within 'Centaur Chess' and robotic surgery, raises questions of the relative ability of each partner and so how labour is divided within the partnership. And that may change over time. Whereas human-AI teams may once have enjoyed a competitive advantage over machines at chess, computers have now developed to the extent that any human involvement only seems to reduce their playing ability.[8] This may also be becoming true of flying a plane. According to a joke amongst pilots, the optimal flight team consists of a human, a dog and a computer.[9] The computer is there to fly the plane, the human is there to feed the dog and the dog is there to bite the human if they touch the computer!

Humans overriding machines can result in a variety of outcomes. One of the greatest stories of a human preventing machine-caused disaster is that of Stanislav Petrov. Stanislav was a Russian military officer in charge of monitoring the nuclear early warning system protecting Soviet airspace during the Cold War. This is how Hannah Fry reports his experience:

> Petrov was on duty on 26 September 1983 when, shortly after midnight, the sirens began to howl. This was the alert that everyone dreaded. Soviet satellites had detected an enemy missile headed for Russian territory. This was the depths of the Cold War, so a strike was certainly plausible, but something gave Petrov pause. He wasn't sure he trusted the algorithm. It had only detected five missiles which seemed an illogically small opening salvo for an American attack.
>
> Petrov froze in his chair. It was down to him: report the alert, and send the world into almost certain nuclear war; or wait, ignoring protocol, knowing that with every second that passed his country's leaders had less time to launch a counter-strike.
>
> Fortunately for all of us, Petrov chose the latter. He had no way of knowing for sure that the alarm had sounded in error, but after 23 minutes - which must have felt like an eternity at the time - when it was clear that no nuclear missiles had landed on

Russian soil, he finally knew he had been correct. The algorithm had made a mistake.[10]

Stanislav Petrov had trusted his human instincts and he brought us back from the brink of nuclear war.

Although not on the scale of nuclear war, thirty-two years later two other people had to decide whether or not to override a machine. They too chose to do so, but on that occasion, it brought injury and pain. At the *Alton Towers* theme park back in 2015 there was an unfortunate incident caused by a fault on a rollercoaster.[11] Two engineers fixed the fault and then sent an empty carriage around the ride to check everything was working correctly. What they didn't notice was that the carriage didn't return. A second error caused it to roll back and to settle in the middle of the track. The machine was built to deal with that and an alarm sounded. But the engineers were confident they'd just fixed the fault so they silenced the alarm and a carriage of cheerful passengers were sent off round the ride, only to crash into the stranded carriage causing injuries to a number of people, including two teenagers who both lost their legs.

If we're going to collaborate effectively with machines, how do we do so well? When do we rely on their judgements? When do we prioritise our own instincts and override the machine? Furthermore, how do we assess human and machine strengths realistically?

The first fatality involving a self-driving car was shocking and quick to be reported, yet we are aware that human driving errors result in serious accidents every single day. In 2015 alone road traffic accidents caused 1.2 million deaths globally.[12] That's not to mention further deaths involving aircraft, trains and boats. To assess machine strength against human ability involves looking realistically at the two, not comparing the former to an idealised standard.

IBM Watson's initial work beyond *Jeopardy* has been in the field of healthcare, but it's possible to imagine it being used in collaboration with other human experts. Partnership with judges within the legal system is one possibility. Would such deployment be wise? Some of us might feel distinctly uncomfortable about machines having an influence on whether or not someone goes to jail and for how long. Nonetheless, as with self-driving cars, our choice is not between a flawed machine and a perfect human system. We know that human systems contain their own imperfections.

A 2012 Israeli study found that the parole judgements given by judges were strongly linked to the time of day in which they were given.[13] The chance of being given parole immediately after the judge had eaten breakfast was 65%. Right before lunch the chance of being given parole had fallen to just 15%. The study has proved contentious and there have been numerous questions around whether or not hunger might have such a significant effect upon human judgements. Nonetheless, whatever the outcomes of those debates and the validity of the study, each of us knows that to be human is to be embodied. Could any of us argue that our own judgements and decisions are completely unaffected by our tiredness, emotions, hunger, hormones, and whether our environment is warm, light or spacious? Might machines assist judges to remove some of the inconsistencies introduced by hunger, distraction and other forms of unhelpful bias?

***

The biblical truth of the Incarnation, of God, in Jesus, becoming a human being, gives further weight to the significance of human beings which we explored through the creation truths of Chapter One, and through the command to love our Creator God in Chapter Three. The affirmation of human beings, made in God's image and created and declared good by Him, is further emphasised by God taking upon Himself that same human form. Jesus' incarnation affirms the significance of what it means to be human. So, when we consider human-machine collaboration, reflection upon this truth of God Himself, in Jesus, becoming human may encourage us to never lightly neglect the significance of what humans can offer within such a partnership.

Holding onto this biblical truth will involve avoiding inappropriate delegation of responsibility to machines. When might such delegation be appropriate? As with each of the biblical truths we have been exploring, they will always be applied most fruitfully alongside one another. Yet the contribution of this particular truth will involve an affirmation of the human. In any proposal to completely delegate human activity to a machine, it might helpfully prompt the question of whether a human-machine partnership might be a preferable alternative. Within a human-machine partnership, it may prompt questions such as: Does this system enhance what might be achieved through human ability alone? Does this partnership delegate tasks according to the strengths of each party? Is there the possibility of a human override? Has the human been well trained to operate any override to best effect?

***

Aubrey de Grey dreams of ending aging.[14]

Global life expectancy has changed dramatically over the last two hundred years.[15] From a common baseline figure of around thirty years in pre-modern times, life expectancy started to increase in early industrialised countries from the beginning of the nineteenth century. In the rest of the world figures stayed low until the last few decades when they then began to catch up. Since 1900, global life expectancy has more than doubled and is now over 70 years. At the time of writing, the country with the lowest life expectancy is the Central African Republic with 54 years.[16] In Hong Kong life expectancy has reached over 85 years. The decline in child mortality and decreasing health inequalities have contributed to these statistical changes.

Aubrey de Grey, aware of these facts, yearns not just for the continuation of this trend but for very much more. It is the upper end of human life expectancy upon which he has set his sights. He is outraged by the idea of aging leading to death and he is committed to working for radical human life extension. He is also optimistic about the outcomes of that work.

Aubrey worked in Information Technology before becoming involved in gerontology, the study of aging, through his wife who is an expert in the field. He believes that we can treat the human body as we do a classic car in order to vastly extend human life expectancy.[17] As we replace the engine parts and body work of a classic car to maintain its function for as long as we wish, so he supposes we can keep on repairing and replacing human body parts too in order to live longer. How long? He believes that there are children alive today who will live to be one thousand years old. Yes, that's right! One thousand years old.[18]

Whilst the basis for Aubrey's hope is somewhat conceptual at this stage, he has described practical outlines for achieving his desired goals. He has identified seven areas which require addressing in order to overcome aging.[19] Each of these represents a very sizeable challenge; the fight against cancer is just one of his seven areas. Nonetheless he is committed to doing what he can, and he expects change to come fast. He's reported as believing that 'the first person to live to be 1,000 will probably only be 10 years younger than the first 150-year-old'.[20]

Aubrey de Grey equates death by aging to death by terrorism[21] and he calls for others to join him[22] in the fight to end what he sees as an intolerable wrong.

***

The fact that Jesus rose from the dead and ascended into heaven in his early thirties or thereabouts means that He didn't become elderly before His earthly body was transformed into a resurrection body.[23] Therefore drawing on Jesus' incarnation within our thinking about aging will need to be slightly indirect. What we do know is that Jesus was born as a human baby and He grew at the pace of human childhood. He didn't need to do that, but He chose to submit Himself to all the inherent limits involved in being fully human. The One who had created the universe was willing to grow and develop, year by year, baby to toddler to child to teenager to adult.

The biblical truth that Jesus grew at the pace of human childhood seems to me to encourage us to embrace the inherent pace and extent of human development, mortality included. That doesn't mean that death has the final say. We'll come on to explore that in the next chapter. But Jesus' willingness to embrace the human form, including the limits and pace of that form, seems to me to encourage each of us to be willing to do the same. If that's correct, then aging is neither to be fought nor to be feared.

As we have already explored, that's not to deny the goodness of healing. We can rightly and appropriately seek to bring healing wherever there is the potential for life. I delight in the fact that medicine now enables the extension of both length and quality of life, even for those who are very old. I see no mandate at all for withdrawing treatment and care from another human being purely on the basis of age. Yet neither, if we affirm the biblical truths of human identity and Jesus' own embrace of this form, do we need to deny the realities of human aging.

These biblical truths enable us to challenge Aubrey de Grey in his failure to distinguish between different causes of death. Terrorist fatalities result from horrific human action. Such deaths are outrageous, avoidable and to be condemned. On the other hand, a peaceful death at the end of a long life, well-lived, while also likely to involve depths of grief, is in its nature entirely different. Human aging and the death it will eventually lead to is an inherent consequence of human growth and development. The fact that Jesus was willing to submit Himself to that same growth and development as a child and young adult is good reason to think twice before seeing those realities as something to be fought at all costs in our own lives and in the lives of those we love.

***

Baroness Susan Greenfield is a neuroscientist who was a previous Director of *The Royal Institution*.[24] Drawing upon her professional experience, in her book *Tomorrow's People* she imagines a future based upon developments in science and medicine.[25] Her vision of the future is one in which people are more and more disconnected from others. Electronic butlers and robotic devices not only respond to our every physical need, but also provide emotional support and companionship. Greenfield questions why anyone in such a world would wish to interact with other humans in such a future:

> If virtual friends replace flesh-and-blood ones, we shall not need to learn social skills, nor think about the unwanted and unpredictable reactions of others. So within this collective consciousness there need be no interaction, no action or response but rather, should we choose it, a passivity in which we are shielded from any disagreement or disharmony. Able to access any information we wish, and capable of choosing from a variety of cyber-companions, what would be the worth in seeking out real-life human individuals? If you were to find them, why would they be interested, or interesting? They would be busy talking with their cyber-friends, or their butler, or watching their favourite film, with, of course, their favourite ending.[26]

***

As we explored in Chapter Four, the command to love our neighbour as ourselves challenges any way of life which makes it less likely that we will be able to do so. The future which Susan Greenfield describes in which we might become so cocooned by machines that we stop interacting with other people risks exactly that. As the command to love our neighbour as ourselves challenges such a lifestyle so too does the fact that Jesus' earthly life was a life lived in community. He modelled for us what it is to be fully alive, and His life involved family, friends and neighbours.

As Jesus' incarnation lends further weight to the significance of the human body it also affirms the value of living in community with other people. If our use of machines withdraws us from positive human interactions, such behaviour may be challenged, not only on the basis of the command to love our neighbour as ourselves, but also on the basis of the relational character of the Incarnation.

***

The introduction of *Flippy*, the burger-flipping robot, into the *Caliburger* restaurant in Los Angeles in March 2018[27] is just one recent example of automation which we can trace back to the industrial revolution and before. Despite *Flippy* being taken offline after just one day due to initial difficulties[28], it was back to work a couple of months later and serving three hundred burgers a day.[29] David Zito, the CEO of *Miso Robots* who created *Flippy*, noted that no one lost their job to *Flippy*, who works alongside human colleagues. Nonetheless, the potential for job losses through these and similar emerging technologies are very real. These potential losses are not restricted to either manual labour or to particular fields of employment. The encroachment of machines upon the workplace is nothing new, but it is expanding.

Back in 2017, a robot priest was deployed into action in Martin Luther's home town of Wittenberg, Germany, to mark 500 years since the Reformation.[30] *BlessU-2* was based in a local church and offered blessings in five different languages. Not many human priests can manage that! As well as being able to select from German, English, French, Spanish or Polish, those wishing to receive a robotic blessing were also able to choose between a male and female voice.

We may not feel that we need to take the idea of electronic priests too seriously, but advances in robotics and artificial intelligence do threaten serious increases in unemployment.[31] One aspect of that threat comes from requiring fewer people to run certain industries. As technology writer and think-tank director Nigel Cameron wrote in 2017:

> The old-style photography leader *Kodak*, which has essentially collapsed, once employed 145,000 people. *Instagram*, today's photography leader, had just thirteen employees when *Facebook* purchased it (*in 2012*) for $1 billion. And to take the most stunning example, the communications sensation *Whatsapp* has been valued at $19 billion. It provides jobs for just fifty-five employees.[32]

Another aspect of the threat of increased unemployment arises from the fact that across our various workplaces, what used to be the exclusive domain of qualified professionals is rapidly becoming open to amateurs equipped with technological support.[33] For example, online medical networks like *WebMD* provide online health information directly to patients, who may then avoid the need to see a doctor. Even back in 2014, the *WebMD* website received 190 million unique visits each month, more than the total number of visits to all the doctors working in the USA.[34]

Similarly, Massive Open Online Courses (MOOCs) enable hundreds of times more students to access courses run by universities such as Harvard and MIT, than are able to attend in person.[35] In the field of law, online dispute resolution (ODR) can avoid the need to go to court. ODR is being used to help resolve 60 million disagreements arising each year between traders on *eBay*, more than three times the total number of lawsuits filed in the entire US court system.[36]

We may appreciate some or all of these developments but whatever our assessment of their value, automation poses a threat, not just to low paid jobs and manual work, but to employment of all varieties.

Many experts predict that we may soon witness the launch of systems of Universal Basic Income (UBI).[37] They argue that as so many humans will be out of work and as wages disappear, in order to keep the economy intact all adults will need to receive an alternative source of provision. Ray Kurzweil expects that in the West all adults will receive UBI by the early-2030s and that such payments will be received globally by the end of that same decade.[38]

How might young people be educated to flourish in all the unknowns of the world of tomorrow? Kevin Roose perceives that in order to succeed in our contemporary technological context, young people have been encouraged to learn technological skills, such as maths, engineering and computer science. Whilst those skills are undeniably important, he argues that it is precisely those fields in which humans are least well placed to compete with machines.[39] He is not the only one to have raised the question of whether, as we train machines to behave like humans, we unthinkingly encourage humans to work more like machines.[40]

Kevin advocates instead prioritising those areas of education within which machines will gain ground most slowly. He suggests those are areas of surprise, care and crisis. Illustrating a context of surprise, he considers the work of a kindergarten teacher and proposes that such a role is still a long way off being performed competently by a machine.[41] In relation to social care, he points to: counselling; religious contexts; nursing; and care of the young, the old and the vulnerable. He suggests that in those arenas we place a high value on the emotional connection available when such care is provided by a fellow human.[42] In the third area, that of crisis, such as when contacting emergency services, Kevin argues that whether or not a machine might competently respond to our need, we greatly prefer the assurance of knowing that a fellow human understands and is providing help.[43]

Kevin Roose also writes of leaving 'handprints'.[44] He describes a handmade ceramic bowl selling for $750, and a new Blu-Ray DVD player for just $47. He notes that:

> The Blu-Ray player was a sophisticated piece of technology with hundreds of parts, assembled by robots in a cutting-edge factory, whereas the bowl was a simple object made out of clay on a wheel, using techniques that are thousands of years old. And yet, the bowl was selling for nearly twenty times the price.[45]

It was the human creativity involved in making the bowl which made the difference – the handprints left upon it. In professional contexts, and throughout our lives, we face the choice of whether to celebrate the characteristics of our humanity or to see them as imperfections to be overcome. Kevin Roose calls us to celebrate those characteristics.

Of course, the future impact of technology on jobs is far from certain. As a child of around eight years old in the late 1970s I remember watching a television programme in school. It portrayed a vision of the not-too-distant future in which technology would bring about shortened working weeks. More than forty years later technological and economic forces seem to have combined, if anything, to do the reverse. The use of email and other communication technologies means that increasing numbers of workers not only face the pressure of work during the day, but also throughout evenings, weekends and even holidays.

As that earlier vision proved to be misguided, so future technological change may not bring about extreme levels of unemployment. We don't know. Yet we can be confident of change, and preparing for scenarios such as extreme unemployment will help us to face whatever future does come about.

**

There are a number of biblical truths which might be brought to bear upon the issue of technology threatening jobs. We might think of creation and explore work as an aspect of life, commenting on the dignity and purpose which it can bring. We might draw upon the command to love one's neighbour as oneself to explore a system of financial welfare to safeguard those unable to gain paid employment. What further illumination might come from considering the third biblical truth of the previous chapter, that Jesus calls us to be fully alive as human beings and that He models for us what it means to do that?

Jesus Himself directly models the reality that work is not limited to paid employment.[46] I imagine that in His early adult life, Jesus generated income through carpentry, using skills taught to Him by Joseph. Yet, Jesus' ministry of healing and teaching described in the gospels was carried out, not for any earthly pay packet, but simply in obedience to the call of His heavenly Father.[47]

Abundant life certainly requires provision. Ensuring that nobody is prevented from enjoying fullness of life will involve thinking through issues of justice in society and the ways in which resources are shared now and in the future. If most wealth in the future is generated by machines, it will be important to consider how that wealth is divided and what social and political structures might need to be put in place to prevent the gap between rich and poor becoming yet greater.

The work of an abundant life, however, is not primarily about provision. The sense of vocation to serve God, whether in education, healthcare, business, church leadership or any other field affirms this reality. It points to activities which have value in and of themselves.

Fullness of life involves purpose and meaning. It involves the opportunity to use the gifts we have been given, to be creative and to be useful. It involves growth, learning and fruitfulness. Play and leisure will be an important aspect of that but whilst the Bible describes the importance of a weekly day of rest, it does so within the context of work.[48] Whatever happens in the new heaven and earth, for now there is always work to be done.

As much of the work that is done today, such as raising children and caring for elderly relatives and friends, does not come through employment but rather through responding to need, so too in the future, we might find that for increasing numbers of us, our 'work' is generated not by the demands of an employer but through our own response to needs around us. When we care for our own bodies through diet and exercise, we do so not because we are paid but in response to need. So too, we face the opportunity of growing increasingly attuned to caring for the needs of others and of the world.

Aspiring for us all to experience the abundant life of which Jesus' speaks affirms the value of using our time for those activities to which humans are so well suited, and also reminds us that as humans we all have our own unique character and make-up. We can join Kevin Roose in celebrating the handprints of our humanity. Work which contributes most to abundant life will be well-fitting. A vocation will emerge out of the overlap between what we are best placed to give and the specific needs around us. It will involve our own human creativity and it is of value no matter how technology develops.

Whether it's as leisure or work, the creation of music, books, plays, films, poems, stories, dance, paintings and other art can bring life and joy both to the creator and to others. It's possible to imagine a future in which a sufficiently capable machine might produce art judged to be of a higher standard than that produced by humans. Yet that in itself will never to be a reason for humans to stop aspiring to produce great art. Work such as the creation of art is about more than the end product. For a human to work to create art is of inherent value. Being fully alive is about aspiring to do all we can well, whether or not machines might do it more efficiently.

Technology may well affect future levels of both income and employment. It may result in machines which are increasingly efficient at carrying out tasks which until now have been the sole preserve of humans. Yet technological development will never remove the opportunity for humans to live life to the full. Our discernment will involve reflecting upon the ways in which technology helps or hinders that possibility.

***

As following Jesus' call to be fully alive as human beings involves work, so there is some work which it is never appropriate for us to shoulder. The Bible describes how the task of bringing about the future has already been accomplished through God's own work. This task is not ours to attempt. It is the topic of our next chapter.

SUMMARY

1. *IBM*'s *Watson*, developed to compete in the quiz show *Jeopardy*, is being used in collaboration with human doctors to help with medical diagnosis and treatment.

2. 'Centaur Chess' is another example of human-machine collaboration as opposed to either full human or full machine control.

3. Stanislav Petrov's experience in 1983 is an example of a human overriding a machine process and so bringing about a profoundly beneficial outcome.

4. The 2015 *Alton Towers* crash provides an example of the problems which may be caused if humans make the wrong call as to when to override a machine process.

5. The Israeli study of judges, and statistics of road traffic accidents, remind us that as machines mimic human activities, the bar currently set by humans is not one of perfection.

6. Jesus being born as a human lends weight to thinking seriously before removing the human element from any human-machine partnership.

7. Aubrey de Grey has a vision of ending aging and believes there are children alive today who will live to be one thousand years old.

8. Jesus growing at the pace of human development seems to caution against any visions which seek to deny the value and meaning of processes of human development and aging.

9. Susan Greenfield imagines a future in which humans become increasingly disconnected from one another through choosing to be enclosed within a technological cocoon.

10. The fact that Jesus' life is one lived in community may lead us to avoid lifestyles and actions which might threaten our connections to other humans.

11. Contemporary technology increases the possibilities for using machines to replace or partner with humans in the workplace, so threatening jobs.

12. The biblical vision of being fully alive offers a way of thinking about using time well in a world of extreme unemployment and preparing children and young people for such a possibility.

QUESTIONS FOR REFLECTION

1. How do you feel about the idea of a robot surgeon, or of machine input into a medical diagnosis?

2. How do you feel about the idea of courts using machines to assist human judges to make wise and consistent decisions?

3. Have you ever had to decide whether or not to override a machine?

4. Would you like to live to be one thousand? Why?

5. Does it make sense to you to equate death by aging with death by terrorism as Aubrey de Grey seems to do?

6. How do you react to the imagined future which Susan Greenfield describes?

7. Would you value a blessing from a robot priest? Why?

8. Has retirement, unemployment, furlough or even holiday led you to wonder how to use your time well? If so, what conclusions have you reached?

1   Steve Jobs, interviewed by Jeff Goodell, *Rolling Stone* Magazine, 16th June, 1994: https://www.rollingstone.com/culture/culture-news/steve-jobs-in-1994-the-rolling-stone-interview-231132/ Accessed 9th Sept. 2021.

2   Stephen Baker, *Final Jeopardy*, p189-209.

3   https://www.mpo-mag.com/contents/view_online-exclusives/2016-08-08/ibms-watson-diagnosed-a-rare-condition-that-left-doctors-stumped/ Accessed 9th Sept. 2021.

4   Garry Kasparov, *Deep Thinking*, p221-248

5   https://mindmatters.ai/2019/08/why-was-ibm-watson-a-flop-in-medicine/ , https://futurism.com/doctors-ibm-watson-ai and https://spectrum.ieee.org/biomedical/diagnostics/how-ibm-watson-overpromised-and-underdelivered-on-ai-health-care All accessed 9th Sept. 2021.

6   Hannah Fry, *Hello World*, p100.

7   Max Tegmark, *Life 3.0*, p101-2.

8   Kevin Roose, *Futureproof*, p20.

9   Hannah Fry, *Hello World*, p22.

10 Ibid., p19-20.

11 Ibid., p20.

12 Max Tegmark, *Life 3.0*, p98.

13 Ibid., p105. Max Tegmark references the original study by Danziger and colleagues, the criticism by Keren Weinshall-Margela and John Shapard, and also the response by Danziger *et al.*

14 Aubrey de Grey, *Ending Aging*.

15 https://ourworldindata.org/life-expectancy#life-expectancy-has-improved-globally Accessed 9th Sept. 2021.

16 https://www.worldometers.info/demographics/life-expectancy/#countries-ranked-by-life-expectancy Accessed 9th Sept. 2021.

17 Aubrey de Grey, *Ending Aging*, p44-45.

18 Ibid., p325.

19 Ibid., p43.

20 https://www.cambridgeindependent.co.uk/business/living-to-1-000-the-man-who-says-science-will-soon-defeat-ageing-9050845/ Accessed 9th Sept. 2021.

21 Aubrey de Grey, *Ending Aging*, p8.

22 Ibid., p311-324.

23 See Chapter Seven for a discussion of Jesus' resurrection body.

24 Baroness Susan Greenfield was Director of *The Royal Institution*, 1998-2010. Back in 1994, she gave the institution's *Christmas Lectures* https://www.rigb.org/christmas-lectures Accessed 9th Sept. 2021.

25 Susan Greenfield, *Tomorrow's People*.

26 Ibid., p43.

27 https://losangeles.cbslocal.com/2018/03/06/burger-flipping-robot-pasadena/ Accessed 9th Sept. 2021.

28 https://www.bbc.co.uk/news/technology-43343956 Accessed 9th Sept. 2021.

29 https://eu.usatoday.com/story/tech/talkingtech/2018/05/28/hamburger-making-robot-flippy-back-serving-300-burgers-day/649370002/ Accessed 9th Sept. 2021.

30 https://www.theguardian.com/technology/2017/may/30/robot-priest-blessu-2-germany-reformation-exhibition Accessed 9th Sept. 2021.

31 Susskind & Susskind, *The Future of the Professions*; Nigel Cameron, *Will Robots take your Job?*

32 Nigel Cameron, Will *Robots take your Job?*, p63.

33 Susskind & Susskind, *The Future of the Professions*, p9-45.

34 Ibid., p47.

35 Ibid., p58.

36 Ibid., p70.

37 Martin Ford, *Architects of Intelligence*, p113, 180, 383, 398, 486.

38 Ibid., p247; https://www.youtube.com/watch?v=XSqzzfhExzw Accessed 9th Sept. 2021.

39 Kevin Roose, *Futureproof*, p61-78.

40 https://www.bbc.co.uk/news/technology-56414491 Accessed 9th Sept. 2021.

41 Kevin Roose, *Futureproof*, p67-69.

42 Ibid., p69-71.

43 Ibid., p71-74.

44 Ibid., p115-130.

45 Ibid., p122.

46 John 5:17b: 'My Father is always at his work to this very day, and I too am working'.

47 John 4:34: 'My food', said Jesus, 'is to do the will of him who sent me and to finish his work'.

48 Exodus 20:8-11: 'Remember the Sabbath day by keeping it holy. Six days you shall labour and do all your work, but the seventh day is a sabbath to the LORD your God. On it you shall not do any work, neither you, nor your son or daughter, nor your male or female servant, nor your animals, nor any foreigner residing in your towns. For in six days the LORD made the heavens and the earth, the sea, and all that is in them, but he rested on the seventh day. Therefore the LORD blessed the Sabbath day and made it holy'.

7

RESURRECTION AND NEW CREATION

But Christ has indeed been raised from the dead, the firstfruits of those who have fallen asleep.
1 Corinthians 15:20

What has been the happiest day of your life so far?

We sometimes hear athletes identify their happiest day after a remarkable sporting success. A student might experience a similar sensation upon graduation. As we look back on our lives, we may name the day we got married, a first date with our spouse, or the days our children were born. We might name the end of a war or a release from captivity. In childhood, we may have a couple of happiest days of our life, as well as three worst days, all within the same week!

What about the most significant day in your life? Or the most significant day in the history of the universe?

I remember as a young teenager having an intense discussion with a friend about which of Christmas and Easter was the more important. We set aside thoughts of chocolate eggs and wrapped gifts and earnestly debated the significance of the two festivals! I started off by arguing that if you could only have one, surely you choose Christmas, when Jesus was born, because if Jesus isn't born then you can't even start to have Easter. My friend attempted to explain to me how it was Easter which was the focal point of the Christian story. I don't remember if he persuaded me at the time but as I look back, I agree with the reality which he was expressing. Of course, we don't need to choose between the two. It's a false choice. Yet the conversation has stuck with me for more than three decades because of the truth my friend helped me understand about the significance of that first Easter weekend.

The Bible points to the death and resurrection of Jesus as the central point of all history. Even our language and the use of the words 'crux' and 'crucial' can remind us of the centrality of Jesus' death upon a cross.

On a Roman cross, an instrument of capital punishment, around two thousand years ago, the Son of God died to accomplish the ultimate rescue of all people. On the cross, Jesus took on Himself all the putrid consequences of our own disobedience, selfishness, laziness and greed. His love, His holiness and His intrinsic goodness destroyed all that stands in the way of us enjoying intimacy with Him, our Creator. That's why the Friday on which all that took place is described as being Good.

In Chapter Three we noted how God rescued His people, Israel, from slavery in Egypt. Echoes of that rescue are found throughout the Bible and, to this day, Jews and Christians around the world mark it annually within Passover celebrations. It was at such a Passover meal two thousand years ago that Jesus met with His friends the evening before His crucifixion. That rescue from Egypt always pointed ahead to the greater rescue which Jesus would go on to accomplish. Out of love for each of us, on the cross, Jesus brought not just freedom from human oppression, but also release for all people from the slavery of sin, the world and the devil. He overcame even the power of death.

How did Jesus do this? Why did His death lead to our rescue? I recognise that we don't need to know everything. The truth of the cross is not a concept to be understood as much as a gift to be received. Nonetheless, the entire Bible is centred on Jesus' death and resurrection and it does provide us with some answers. This is how it is expressed in the book of 1 Peter: '"He himself bore our sins" in his body on the cross, so that we might die to sins and live for righteousness; "by his wounds you have been healed"'.[1]

Through a process that theologians call atonement, Jesus took our place when our selfishness threatened to separate us from enjoying intimacy with the loving purity and holiness of our Creator God. Jesus, who was completely without that selfishness, paid the price for ours, and so enabled us to come close to Him and to enjoy His friendship.

I am aware that unlike Jesus, Maximillian Kolbe was neither God nor perfect. Nonetheless, I think Kolbe's cross-inspired sacrifice offers an image of how Jesus set us free. On 17 February 1941, this forty-seven-year-old follower of Jesus was arrested by the Gestapo. The monastery in which he served was shut down by the authorities on the same day. Just over three months later, on 28 May, Kolbe was transferred to the concentration camp at Auschwitz. At the end of July that year, when one prisoner managed to escape from the camp, the deputy camp commander ordered that ten prisoners be starved to death in order to deter further escape attempts. One of those chosen, Franciszek Gajowniczek, cried out, "My wife! My children!" Kolbe, without a wife or children, volunteered to take his place. The authorities accepted Kolbe's request and over the following days

Kolbe led his fellow prisoners in prayers as they were deprived of both food and water. By 14th August 1941, of the ten prisoners, only Kolbe was left alive. The guards then killed him by lethal injection in order to empty the bunker in which he and the others had been imprisoned.

Kolbe's sacrifice was his way of expressing the love He'd already received from Jesus. As such, his actions, made possible through the Holy Spirit at work within him,[2] point towards Jesus' own sacrifice and give us a glimpse of what it meant for Jesus to die for us upon the cross.

***

The rescue of us all by Jesus upon the cross is inextricably connected to His resurrection from the dead. It is because Jesus is fully God as well as fully human that His death upon the cross made a difference to my relationship with God. Because Jesus is God, His sacrifice of Himself wipes away all that would otherwise separate us from Him. For the same reason, precisely because Jesus is God, death could not hold Him. As the Old Testament had prophesied, and as the New Testament declares, Jesus rose again from the dead on the first ever Easter morning.[3]

There's joy-filled surprise at the heart of Easter. Jesus' followers weren't expecting Him to rise from the dead! Even though He'd been preparing them for all that was about to happen, the enormity of what God was about to do was too much for them to grasp. It was only looking back afterwards that they were able to see how Jesus had been heading this way all along.

I was involved in an evening event with a group of teenagers some years ago and a mime artist, who went on to be the Best Man at my wedding sometime later, was performing a drama. He was acting out Jesus' death upon the cross and at the point of death, with arms outstretched, the background music fell silent, his head dropped, and all was still. The watching teenagers waited for a moment and then applauded as they showed appreciation for the power of the drama which they felt had reached its end.

Suddenly the *Chumbawamba* song *Tubthumping* leapt out through the PA system. The actor, previously in crucifixion pose, looked up and winked, as the lyric 'I get knocked down but I get up again' rang out around the marquee. The teenagers were thrilled and delighted and erupted into a second and richer round of applause, this time accompanied by laughter, as the joy-filled surprise at the heart of Easter had been conveyed afresh.

There are many things which Jesus' resurrection from the dead was not. Amongst them, this was not a hoax, nor an allegory, nor a resuscitation.

This was not a fiction invented by the disciples at the time, nor was it an allegorical device used in later years to shape a movement by poetically expressing vague religious concepts which never really took place. Jesus' resurrection was both physical and historic. It was also a doorway to new life, not a resuscitated return to an old one.

By resuscitation I am referring to any event looking like death, from which someone returns to life, only to go on and to die at a future point. We might discuss, in different cases, whether or not death occurred at all before the point of resuscitation. What is certain is that death went on to take place, or will go on to take place, after the point of resuscitation.[4] The raising of Lazarus from the dead by Jesus is one such event.[5] The Bible makes clear that in that case, death had already occurred.[6] Jesus brought Lazarus back from the dead to enjoy extended life this side of death. Jesus did not, at that time, take Lazarus beyond death, immediately into resurrection life. Lazarus went on to die again, at a later point, and only after that did the possibility of resurrection life for Lazarus become real. Resuscitation also covers all cases of near-death experiences even after a medical pronouncement of death. In all these cases eventual death is postponed rather than passed beyond.

When a child is born, medical interventions might speed up or slow down the exact moment of birth. What is clear is that life inside the womb is fundamentally different to life outside. Resurrection represents a similarly fundamental boundary. Resurrection life comes beyond death in such a way that death is no longer a possibility. Resurrection life is to resuscitation what childhood development is to the moment of birth being delayed.

The Bible gives us some clues as to how Jesus' resurrection life was different to His life before. After His resurrection Jesus appeared to His followers on a number of occasions.[7] His body at that time, His resurrection body as the Bible calls it, was a matter of interest to the gospel writers and others. That body was both like and unlike his previous body. It pointed ahead to the new life which He had brought about. After the resurrection, Jesus was recognizable but often not easily so.[8] He ate fish and yet could pass through locked doors.[9] He invited Thomas to touch Him, yet told Mary not to hold onto Him.[10]

The Bible tells us that Jesus' own embodiment after the resurrection is a foretaste, a 'firstfruit' as the New Testament puts it, of our own embodied existence beyond death.[11] It points, not to an immaterial, ghostly, disembodied future but to physical resurrection bodies, like that of Jesus Himself.[12]

After these resurrection appearances, the Bible describes how Jesus then ascended into heaven.[13] His followers watched as He rose up and angelic figures assured them that He would one day return in the same way. Shortly afterwards, as Jesus had promised, [14] the Holy Spirit was poured out upon those disciples, a turning point marking the invitation for all Jesus' followers to be filled in a similar way.[15] As God had accomplished what He needed to through having all His fullness dwell in human form, He now moved on to the next stage of His good plans and purposes, becoming particularly present with all those who wished to open themselves to His presence, empowering, strengthening, guiding and equipping.

***

And so, the work of New Creation began. That New Creation is both on its way and not yet fully here. We catch glimpses of it as we await its ultimate arrival.

We live at a point in history between the cross and the resurrection of Jesus on one hand, and a new heaven and a new earth on the other, between Jesus' first and second comings. We await His return and His ushering in of the New Creation. Jesus has already secured and achieved the final victory through which death and evil will be destroyed and New Creation will be revealed. He accomplished that through His death and resurrection. The reality of that assured victory is testified to throughout the Bible, not least in the complex and potentially confusing biblical book of Revelation which describes it repeatedly, from different perspectives, using a rich variety of imagery.[16] New Creation is already assured.

The Bible seems to point towards some sort of continuity between the New Creation and the Old. As the Bible does not describe a future disembodied existence neither does it portray an end to God's plans for earth. It points to a new earth as well as a new heaven.[17]

We can only imagine the detail of that renewed reality[18] but the Bible assures us that it will be a place of intimacy with God, a place without tears, or pain, or sorrow, or death:

> Then I saw 'a new heaven and a new earth,' for the first heaven and the first earth had passed away, and there was no longer any sea … And I heard a loud voice from the throne saying, 'Look! God's dwelling place is now among the people, and he will dwell with them. They will be his people, and God himself will be with them and be their God. 'He will wipe every tear from their eyes. There will be no more death' or mourning or crying or pain, for the old order of things has passed away.' He who was seated on the throne said, 'I am making everything new!'[19]

Through the cross and the resurrection Jesus has flung wide open the gates of His New Creation to all of us who wish to enter into intimacy with God. Yet, for now, we live in the now and the not yet of His Kingdom. The outworking of Jesus' victory is not yet complete. At this time, we experience viruses and dictators, natural disasters and aging. Such trials are challenging but God gives us the strength with which to endure them. As a biblical writer puts it: 'we are hard pressed on every side, but not crushed; perplexed but not in despair; persecuted but not abandoned; struck down but not destroyed'.[20]

Jesus has released us to follow Him as He unfolds His plans to bring about the future. In chapters Nine and Ten we will explore what it means for us to live as Jesus' followers in the midst of the unfolding of those plans. First of all, in the next chapter, we will explore how that assured promise of the future releases us to live with hope in our technological world. The future is not on our shoulders. We do not bear the burden of bringing it about. As followers of Jesus, we can witness to the freedom to live without the need to secure our own future.

SUMMARY

1. Biblical Truth Eleven: Out of love for each of us, Jesus died and then rose again, not avoiding death but going through it.

2. Through Jesus' death and resurrection our sins have been forgiven and we are enabled to participate in the New Creation which Jesus has already secured.

3. Biblical Truth Twelve: Jesus' body was transformed, after His resurrection from the dead, into His 'resurrection body'.

4. Jesus' resurrection body points ahead to the invitation for us too to receive our own resurrection bodies beyond death.

5. As New Creation involves an embodied future, so too it involves a new earth and new heaven.

6. We live at the time between the cross and resurrection of Jesus, and His future return ushering in the arrival of the new earth and the new heaven.

7. Biblical Truth Thirteen: Jesus' death and resurrection have already secured the future.

8. As the future is already assured, we are released to follow Jesus without the pressure of needing to secure the future for ourselves.

QUESTIONS FOR REFLECTION

1. What would you say about why Jesus died on the cross?

2. How do you feel about the impact of that event upon your own life?

3. How do you feel about the idea of embodied life beyond death?

4. How might a new earth be different to our present earth?

5. Why might we need both a new earth and a new heaven?

1   1 Peter 2:24.

2   Acts 1:8: 'But you will receive power when the Holy Spirit comes on you; and you will be my witnesses in Jerusalem, and in all Judea and Samaria, and to the ends of the earth.'

3   Matthew 28; Mark 16; Luke 24; John 20.

4   Unless the New Creation breaks in first, of course, but that's getting ahead of ourselves.

5   John 11:1-44.

6   John 11:14.

7   1 Corinthians 15:6-8.

8   Luke 24:13-35.

9   Luke 24:36-49; John 20:19-23.

10 John 20:27: 'Then he said to Thomas, "Put your finger here; see my hands. Reach out your hand and put it into my side. Stop doubting and believe."' John 20:17: 'Jesus said, "Do not hold on to me, for I have not yet ascended to the Father. Go instead to my brothers and tell them, 'I am ascending to my Father and your Father, to my God and your God.'"'

11 1 Corinthians 15:35-58.

12 John W. Cooper addresses this issue in *Body, Soul & Life Everlasting* as he explores biblical anthropology and the Monism-Dualism debate.

13 Acts 1:1-11.

14 John 16:7: 'But very truly I tell you, it is for your good that I am going away. Unless I go away, the Advocate will not come to you; but if I go, I will send him to you'.

15 Acts 2:1-13.

16 Tom Wright's book *Revelation for Everyone* may be a helpful place to begin for anyone who desires to explore more about the biblical book of Revelation, the final book of the Bible.

17 Revelation 21:1-5.

18 C S Lewis does so in the final chapters of *The Last Battle*, the last in the seven books making up the Narnia series.

19 Rev 21:1-5a.

20 2 Corinthians 4:8-9.

8

A LIFE WITH A FUTURE

Strength for today, and bright hope for tomorrow …
Thomas Chisholm[1]

The biblical truths from the previous chapter were:

---

11. Out of love for each of us, Jesus died and then rose again, not avoiding death but going through it.

12. Jesus' body was transformed, after His resurrection from the dead, into His 'resurrection body'.

13. Jesus' death and resurrection have already secured the future.

As we turn now to our contemporary context, we begin with a story of a dragon.

***

Nick Bostrom shares with Aubrey de Grey the vision of aging as a problem to be solved by human efforts. Nick is a founder of the *World Transhumanist Association*, now known as *Humanity Plus*[2]. Transhumanists desire to become posthuman: beings with vastly greater capabilities than humans currently enjoy.[3] That quest involves seeking to defeat aging. Nick has written a myth about doing so, using the analogy of slaying a dragon-tyrant.[4] This is how he begins his tale:

> Once upon a time, the planet was tyrannized by a giant dragon. The dragon stood taller than the largest cathedral, and it was covered with thick black scales. Its red eyes glowed with hate, and from its terrible jaws flowed an incessant stream of evil-smelling yellowish-green slime. It demanded from humankind a blood-curdling tribute: to satisfy its enormous appetite, ten thousand men and women had to be delivered every evening at the onset of dark to the foot of the mountain where the dragon-tyrant lived. Sometimes the dragon would devour these unfortunate souls upon arrival; sometimes again it would lock them up in the mountain where they would wither away for months or years before eventually being consumed.

The moral at the end of the tale contains these words:

> Stories about aging have traditionally focused on the need for graceful accommodation. The recommended solution to diminishing vigour and impending death was resignation coupled with an effort to achieve closure in practical affairs and personal relationships. Given that nothing could be done to prevent or retard aging, this focus made sense. Rather than fretting about the inevitable, one could aim for peace of mind.
>
> Today we face a different situation. While we still lack effective and acceptable means for slowing the aging process, we can identify research directions that might lead to the development of such means in the foreseeable future.

"Deathist" stories and ideologies, which counsel passive acceptance, are no longer harmless sources of consolation. They are fatal barriers to urgently needed action.

For Nick Bostrom, aging is just one more item to cross off the human 'To Do' list.

***

This story of defeating aging through human activity is illuminated by the biblical truth of the previous chapter, that Jesus died and then rose again for us, not avoiding death but going through it. We explored in Chapter Six how Jesus' willingness to become human and to develop at the pace of human development challenges any vision of seeing human development and aging as a problem to be overcome by our own efforts. Jesus' willingness to die upon the cross removes any lingering possibility that human aging and death need to be resisted today at all costs.

In ongoing small ways, we are indeed called to fight death. Chapter Two has already explored the importance of medicine and of working for life and health. As I write in mid-2021, I delight in the effects of vaccination and other actions which are reducing daily deaths due to coronavirus across the world. I celebrate all that can be done to fight disease.

Similarly, I am not indifferent to aging. I feel no compulsion to take pleasure in the small but numerous signs that my middle-aged body is not what it once was. I am not bothered by the reality that, little by little, more of my hair is becoming grey but I do miss the strength I once had in my back and the ability of my younger body to heal quickly from cuts and scrapes. I know that coming decades, if I live to see them, are likely to lead me towards an increasing number of the physical and mental 'complaints' and complications which come with age.

I am convinced that it's right for us to care for our bodies through appropriate diet and exercise, recognising the awesome gifts which they are. I am delighted by the way that developments in medicine and society are improving the quality of life in old age, as well as its extent. I am grateful to all who are using their professional lives to work for these aims. Yet I do not wish to fight either aging or death at all costs.

I do not hope to live to be a thousand. I believe that whatever Nick Bostrom may argue, there is still a place for 'graceful accommodation' when it comes to facing both aging and death. I believe that unless the new heaven and the new earth arrive first, our own futures lie beyond death. That's okay. Jesus' victory over death on the cross means that the sting of death has already been removed. There is thus no need for anyone either to live in fear of aging or death, or to see them as ultimate foes yet to be vanquished.

***

Ray Kurzweil's vision of the future is strikingly bold. Yet, despite its scale, it comes with a powerful endorsement. Bill Gates, whose credentials in technology are well-established, describes Ray as 'the best person I know at predicting the future of artificial intelligence'.[5]

Ray Kurzweil predicts the appearance of a 'singularity'.[6] This singularity will be a technological discontinuity, a jolting change of such significance that it will be as if we have passed from one world to another. It will arrive as a consequence of technological change having reached such a level that it is virtually impossible to imagine what life will be like beyond that threshold. In other words, all our known reference points will be of little help in navigating that completely unfamiliar technological context.

Imagine a young woman, a Celtic farmer, a Pict at the time of Hadrian, suddenly transported almost two thousand years into the future and finding herself in the city centre of Edinburgh in the present day. The culture shock she might experience and the technological change she would encounter are similar to the sense of disorientation that Ray Kurzweil predicts beyond the singularity.

Ray describes a world in which it will be unrealistic to expect to distinguish between people and robots. What's more, even the attempt to do so would be to miss the significance of the previous merging of the two. This is how he portrays that time:

> The Singularity will represent the culmination of the merger of our biological thinking and existence with our technology, resulting in a world that is still human but that transcends our biological roots. There will be no distinction post-Singularity, between human and machine or between physical and virtual reality.[7]

Far beyond the internet of things: fridges that remind us of use by dates, and heating systems which respond to news of our unexpected imminent arrival home, Ray pictures a future in which we will be so immersed in our technology that we will never be free to extricate ourselves from it. He imagines our having unified our very selves with technology in such a way that it will not make sense to speak of a boundary between one and the other.

What's more, he is not expressing a vision of a distant future, far beyond our own experience. This is not *Buck Rogers* four centuries from now![8] Ray predicts this singularity will occur within the next generation, by the mid-2040s.[9] Yes, he describes a future in which, by the time that children born today are in their mid-twenties, the world will be unrecognisable. As such a world offers technological possibilities few of us even dream of today, he is committed to being around himself to reap their rewards.

Ray Kurzweil predicts that through caring effectively for his own body until that point, he will be able to achieve a form of immortality through uploading himself into a computer system.[10] He will then be able to move at will within both the real world and virtual reality and to make use of both physical bodies and electronic forms as he chooses.[11]

It may be worth pausing for a moment to reflect upon this astonishing belief. Ray was born on 12th February 1948. During 2045, the year that he sets for the singularity, if he is still alive, he will turn ninety-seven years old. Yet he plans not only to be alive at that point, but to journey further on a technological pathway that will enable him to avoid death for as long as he may desire. Ray is not just dreaming of such an eventuality. He has been actively planning for it over many years.

Ray Kurzweil has invested a great deal of study in learning about the dietary and exercise habits which will maximise his chances of staying alive and healthy for a further three decades. His aim is to place himself in the best possible position for taking advantage of future medical opportunities for further life extension, and eventually for immortality. With his doctor, he has written two substantial books on diet and exercise as part of this process.[12] Ray also ensures that he implements the conclusions of his research within his day-to-day life. He has confirmed taking about 250 pills a day,[13] containing various supplements to his diet.[14] He receives a detoxification treatment each and every week and undergoes frequent and regular monitoring of various physiological measures.[15] Ray is both passionate about, and committed to, his vision for the future.

***

We have already explored the distinction between healing and enhancement. Medicine which seeks to repair the human body is carrying out a healing purpose for which there is biblical affirmation. Such healing repairs may be shockingly profound, such as the transplant of a human face following severe burns. It is not the extent of the medical treatment which determines whether or not it serves a healing purpose. Yet if a medical procedure goes beyond healing and seeks to 'improve' upon our bodies, we might question any such aims on the basis of the biblical truths of the goodness of creation, of loving God, and of Jesus' incarnation, from Chapters One, Three and Five respectively.

The biblical truth from Chapter Seven that Jesus' body was transformed, after His resurrection from the dead, into His 'resurrection body' further illuminates any radical reimagining of our bodies this side of the grave. The future which Ray Kurzweil describes of uploading his body into a computer system and switching between real and virtual bodies at will may be imaginative but it is to deny the goodness of God's own vision of human embodiment. The Bible offers us a vision of human bodies, fearfully and wonderfully made, lasting through life and then transformed beyond death into resurrection bodies.

The hope of resurrection embodiment provides further reason to embrace our present bodies and to resist the desire to overcome their inherent limitations at this point in time. It may prompt us to be more patient with our present embodied struggles knowing that there is already a transformed future awaiting us.

Between clear biblical examples of healing such as curing leprosy and enabling the lame to walk, and Ray Kurzweil's enhancement desire to switch bodies at will, we find more complex ethical territory. It is here that it is harder to identify where the boundaries lie between healing and enhancement. For example, if a person feels uneasy in their own body, is that an illness which requires treatment and if so, how? What some will understand to be a healing procedure will be seen by others as a seeking after bodily 'improvement'. There will continue to be different perspectives. Yet the biblical hope of a resurrection body may provide for some the encouragement to persist for a little longer with an embodied experience that leaves a yearning for more.

A yearning for more may stem from the fact that I need healing. On the other hand, it may reveal the reality that I was made for more, not now but beyond death. My current embodiment is only a taste of what is yet to come. That knowledge may provide the peace to live with uncomfortable aspects of embodied life with the reassurance that resurrection bodies await. It certainly offers a vision of embodied freedom which provides an alternative to Ray Kurzweil's dream of switching physical and electronic bodies at will.

***

By the age of nineteen Danielle has not only been elected chairman of the Communist Party of China but she has also started to lead the nation towards democracy. She has become the first democratically elected president of China. She has received three *Nobel* prizes. That is not to mention the fact that she has also been elected President of the United States.

Danielle is the fictitious heroine of Ray Kurzweil's novel, *Danielle: Chronicles of a Superheroine*.[16] At the age of six she is working with the Zambian government to provide clean water.[17] By age eleven she has received the Nobel Peace prize for bringing peace to the Middle East.[18] At age twelve she develops a cancer treatment that brings remission to ninety-eight percent of all patients.[19] She then goes on to transform China and to win another couple of *Nobel* prizes before turning her attention to the US.

Central to Kurzweil's novel is the dream of achieving utopia through human activity. There's a lot in common with the 'new human agenda' of Yuval Noah Harari referred to in Chapter Two. These visions predate our contemporary technological context. They are as old as humanity. The Bible reveals them to us at Babel[20] and legends such as that of Icarus testify to them too. Yet the technological possibilities of today and tomorrow can suggest tantalising new means of their fulfilment. Writing fourteen years earlier, Kurzweil articulates the centrality of this quest in his own life:

> To this day, I remain convinced of this basic philosophy: no matter what quandaries we face – business problems, health issues, relationship difficulties, as well as the great scientific, social and cultural challenges of our time – there is an idea that can enable us to prevail. Furthermore, we can find that idea. And when we find it, we need to implement it. My life has been shaped by this imperative. The power of an idea – this is itself an idea. ... I recall my grandfather ... (*and*) ... a rare opportunity he had been given to touch with his own hands some original

135

manuscripts of Leonardo da Vinci. ... He described the experience with reverence, as if he had touched the work of God himself. This, then, was the religion that I was raised with: veneration for human creativity and the power of ideas.[21]

Danielle symbolises the dream of overcoming all problems through human creativity resourced by technology.

***

It would be easy to mock the implausibility of Kurzweil's plot within his novel about Danielle but that would be to miss the point of the book. He is deliberately provoking the imagination and stretching our sense of what might be possible. I too believe in a future which might easily be dismissed as being too good to be true. Where I part company from Kurzweil is in the source of such a hope-filled future.

I agree that profound ideas are transformational. I too am inspired by the creativity and imagination of great artists. Yet for me, these symbols of human achievement point beyond ourselves to the infinitely greater wonder and creativity of the One who made us. I cannot join Ray Kurzweil in venerating human creativity and the power of ideas. Not after centuries of war, oppression and discrimination. Not after slavery, abuse and cruelty. Not after the Holocaust. Not after Adam and Eve proved their inability to live a life of abundance within God's good boundaries when presented with the opportunity do so.

The singer-songwriter Bruce Cockburn explores the nature of humanity within his song *Burden of the Angel/Beast*. Self-awareness gives each of us the opportunity to recognise within ourselves the potential for both good and ill which that title expresses. As humans, we can reach inspiring pinnacles. Yet, we can also plunge the depths of depravity and destruction. Some say they struggle to believe in a God of love. For me, the universe itself points to His presence and to His goodness with both jubilant shouts and restrained whispers. How much harder it seems to me to place one's hope in humanity with all our chequered past.

It seems appropriate to me that Ray Kurzweil names religion towards the end of the earlier quote: 'This, then, was the religion that I was raised with: veneration for human creativity and the power of ideas'.[22] The way we think about human nature and potential is completely dependent upon our religious understanding of what it means to be human. Who are we? Where have we come from? Where are we going? What are we here for?

The biblical assurance of the previous chapter is that Jesus' death and resurrection have already secured the future. That truth can shape the way we go on to answer the questions above. It can bring freedom to live as creatures, humans with the God-given potential for wonderful creativity and contribution, but creatures nonetheless. Without the assurance of the previous chapter, we will answer those questions very differently. We may feel as if the responsibility to bring about the future is on our own shoulders. Kurzweil seems to enjoy the idea of such a challenge. Others may feel less comfortable with that expectation.

The biblical promise of New Creation releases us from the need to shape the future for ourselves. Of course, we change the future. We act in ways which will either help or hinder ourselves, others and the planet. These actions matter. Yet the ultimate future is not upon our shoulders. Jesus has already secured that through His death on the cross and His resurrection. He has liberated us to live without fear.

137

He has freed us to live as His followers, as we look ahead to what He has already achieved. How we do that is the subject of our final two chapters.

SUMMARY

1. Nick Bostrom has expressed the transhumanist dream of defeating aging through human efforts.

2. The fact that Jesus did not avoid death but went through it, and in so doing removed its sting, means that we do not need to encounter death or aging as ultimate foes, or as yet to be defeated.

3. Ray Kurzweil believes that technological developments will lead to a singularity in the mid-2040s, providing opportunities to upload ourselves into a computer system and to inhabit real or virtual bodies at will.

4. Ray Kurzweil's vision of uploading oneself and then inhabiting virtual or real bodies this side of death denies both the goodness of human embodiment as created by God and of resurrection bodies beyond death.

5. In his writing of Danielle, and in other ways, Ray Kurzweil expresses the vision of bringing about the future through human activity.

6. The biblical promise of New Creation is of a future brought about not through human activity but through the cross and resurrection of Jesus Christ.

7. Living as a follower of Jesus involves both recognising that our actions have consequences and that those consequences have already been dealt with by Jesus. Our actions matter but the future is secure.

QUESTIONS FOR REFLECTION

1. How do you respond to the concept of Bostrom's tale of the dragon-tyrant?

2. How do you feel about Kurzweil's assertion that the world will be unrecognisable by the mid-2040s?

3. If you could ask the fictitious Danielle one question, what would it be?

4. How do you respond to Kurzweil's comment that for every quandary there is an idea which can enable us to prevail?

1   Within his 1923 poem which was the basis of the hymn *Great is Thy Faithfulness*.

2   https://humanityplus.org/ Accessed 9th Sept. 2021.

3   https://www.nickbostrom.com/tra/values.html Accessed 9th Sept. 2021.

4   https://www.nickbostrom.com/fable/dragon.html Accessed 9th Sept. 2021.

5   Ray Kurzweil, *The Singularity is Near*, piii.

6   Ibid.

7   Ibid., p9.

8   *Buck Rogers in the 25th Century* was a television series running from 1979 to 1981.

9   Ray Kurzweil, *The Singularity is Near*, p136.

10 Ray Kurzweil & Terry Grossman, *Transcend*.

11 Ray Kurzweil, *The Singularity is Near*, p324-5.

12 Ray Kurzweil & Terry Grossman, *Fantastic Voyage*; Ray Kurzweil & Terry Grossman, *TRANSCEND*.

13 Ray Kurzweil & Terry Grossman, *Fantastic Voyage*, p141.

14 https://transcend.me/blogs/supplementation/what-supplements-does-ray-kurzweil-take-and-why Accessed 9th Sept. 2021.

15 Ray Kurzweil & Terry Grossman, *Fantastic Voyage*, p139-145.

16 Ray Kurzweil, *Danielle: Chronicles of a Superheroine*.

17 Ibid., p37-59.

18 Ibid., p123-141.

19 Ibid., p145-166.

20 Genesis 11:1-9.

21 Ray Kurzweil, *The Singularity is Near*, p2.

22 Ibid., p2.

9

THE CHURCH

'And I tell you that you are Peter, and on this rock I will build my church, and the gates of Hades will not overcome it'.
Matthew 16:18

Today there is a cross in the Colosseum.

In the time of the early church, the Colosseum in Rome was the site of the death of numerous Christian martyrs. Tradition has it that Simon Peter, one of Jesus' twelve disciples, was martyred nearby, in Nero's Circus. I wonder what he was thinking as the time of his execution drew near.

Some years earlier, at Caesarea Philippi, Peter had declared that Jesus was the Messiah, the Son of the living God.[1] In response, Jesus had made a promise. He promised that on this rock, He would build His church.[2] Perhaps Jesus meant the rock of Peter himself, or maybe Jesus meant the rock of the declaration that Peter had just made.[3] Either way, Jesus' promise was clear: He would build His church. There were times when Peter will have been encouraged as He saw Jesus bring about the fulfilment of that promise. At Pentecost, as Peter preached, and 3000 people became followers of Jesus on one day,[4] I guess that Peter was feeling pretty good. Awaiting execution in Rome, at the hand of Nero, knowing that Christians were routinely being fed to the lions for the entertainment of the crowds, I can imagine finding it harder to hang onto the truth of Jesus' words.

Yet today there is a cross in the Colosseum. Jesus' words have proven true. His promise continues to be fulfilled. The Roman Empire is long gone but there are local churches in every corner of the globe. Whilst Nero's leadership is finished, the work of which Peter was a part endures and grows. Jesus has been faithful, and He will continue to be faithful, to His promise to build His Church.

During the days immediately following Jesus' resurrection, when His followers were huddled together in fear of the authorities, Jesus appeared to His disciples, He nurtured their faith, and He commissioned them.[5] Following His ascension into heaven, He poured out His Holy Spirit upon them, and they were transformed from timid to courageous.[6] When they were imprisoned by the authorities they were released[7] or their imprisonment was used to work for good.[8] When the temple in Jerusalem was destroyed in AD70, that tragic event was nonetheless used by God to send out His people in witness and in mission to lands that had never before heard the Good News of Jesus. In all these ways Jesus was being faithful to His promise to build the Church.

As followers of Jesus have responded to His call over the last two millennia, the Church has been built in towns, villages and cities. The Church has been present locally in homes and cathedrals, megachurches and prayer triplets. People groups throughout the world have had the Bible translated into their own native tongue. Whilst conflict and dispute between Jesus' followers has led to the sorrow of splits between communities and the formation of new denominations, nonetheless, Jesus has been faithful to His promise to build the Church. With a myriad of worship styles, in buildings of
144

all types and none, with differences in leadership, structure and dress, Jesus' followers have known themselves built up by Him.

Persecution throughout the centuries has been unable to prevent the growth of the Church. Indeed, the blood of the martyrs has even been described as the seed of the church.[9] Jesus has built His Church as His followers have come from every background and lifestyle, from all countries and cultures, from each political system and economic circumstance.

The Bible affirms that 'neither death nor life, neither angels nor demons, neither the present nor the future, nor any powers, neither height nor depth, nor anything else in all creation, will be able to separate us from the love of God that is in Christ Jesus our Lord.'[10] As Jesus continues to pour out His love in the world, so He continues to build His Church.

As we explored in the previous two chapters we live at a time between the resurrection of Jesus in the past and His bringing about of the New Creation in the future.

We might look around us in the early twenty-first century and describe this as the time of technology, or of capitalism, or of democracy, or the time for environmental action, but for followers of Jesus, primarily this is the time of the Church. This is the time when we are called to help establish, point to, and live in the Kingdom of God. We may use technology to do that but we won't serve technology. We will care for the environment but that won't be our sole focus. We will live knowing that our actions have consequences but we won't live as if the fate of the world is in our hands. For this is the time of the Church.

The time of the Church is the period of history in which the Holy Spirit is being poured out upon all peoples. It runs from just after Jesus' ascension into heaven, until His future return[11] and the arrival of the new earth and the new heaven. Within this time, Jesus' followers, empowered by the Holy Spirit, are to spread the message of His good news to the corners of the world.[12] We are to speak of the victory of His cross and resurrection. We are to heal the sick in His name,[13] to pray continually,[14] to call the world to repentance,[15] to baptise in His name,[16] and to trust in the outworking of God's perfect plans.[17]

Our contemporary context prompts myriad questions. What does it mean to engage well with technology? Is it too late to protect our planet? Which species will be alive in the time of our grandchildren? What does it mean to be human in a world of AI? How do I live with hope in a world which prompts me to fear?

We have already explored how creation; God's law; the Incarnation; the death and resurrection of Jesus, and the promise of New Creation all offer insights for our engagement with technology. So too does reflection upon living in the time of the Church.

Two thousand years ago Roman Emperors were confident in the power of their empire. The New Testament bore witness to the reality that the empire's power was limited.[18] History has proved the truthfulness of that biblical testimony. Today other forces assert their own power in turn. They range from tech billionaires to ideological movements, and from nation states to terrorist factions. They too wield only limited power. The Bible reveals that God is faithful in working out His good plans and purposes and that nothing can stand in the way of Him doing so.[19] Followers of Jesus don't need to be overwhelmed by the powerful forces around us. We don't even need to aspire to extra power for ourselves. The One who is the source of

146

all power has assured us that His strength is made perfect in weakness.[20]

Recognising that this is the time of the Church shifts the focus of our attention from the challenges around us to the call to follow Jesus at this time. That doesn't mean those challenges become any less important but it does mean they gain perspective through being placed in a new context. The knowledge that Jesus has already given us all we need to be obedient to His call releases us to question what we do and don't attempt to carry ourselves. If we are unable to solve all the problems of the world that's because He has not called us to do so. He has already accomplished that task Himself. We are free simply, in obedience to His call, to be the Church.

How to be obedient to Jesus' call to follow Him as members of the Church is the adventure of a lifetime. It involves being continually filled with the Holy Spirit,[21] prayer,[22] feeding on the Bible,[23] being joined to our brothers and sisters within His body,[24] and keeping our eyes fixed upon Jesus[25] in order to follow His leading.[26] The next chapter will not provide the answers for how each of us needs to do that day by day, but it will seek to offer a few thoughts about how our contemporary context in the early twenty-first century shapes that call.

SUMMARY

1. Biblical Truth Fourteen: Jesus has been and will always be faithful to His promise to build His Church.

2. We may be tempted to consider our 'now' as the time of technology or the age of AI, yet for followers of Jesus, this is the time of the Church.

3. Biblical Truth Fifteen: Living as a follower of Jesus involves being continually filled with the Holy Spirit, prayer, feeding on the Bible, being joined to our brothers and sisters within His body, the Church, and keeping our eyes fixed upon Jesus in order to follow His leading.

4. Our contemporary context gives a particular flavour to the way in which we obey God's call within our generation.

QUESTIONS FOR REFLECTION

1. What do you think Peter would feel about the Church if he could visit earth today?

2. How do people describe contemporary times, in expressions such as *The Age of ...*?

3. If you are following Jesus, how are you growing in obedience to that call?

4. If you are not currently following Jesus, would you like to do so? Your local Christian church provides a way to join with others who are seeking to follow Him. The Alpha course offers a wonderful way to explore what it might mean to do so.[27]

1   Matthew 16:13-16; Mark 8:27-30; Luke 9:18-20.

2   Matthew 16:17-20.

3   There's certainly word play going on here because the Greek word for Peter means rock. Much teaching points to Peter himself as the rock on which the Church is built. Andy Stanley, in a *Global Leadership Summit* talk from 2013, makes a convincing case for the rock being instead the declaration which Peter has just made, that Jesus is the Messiah, the Son of the living God. He argues that it is on this truth that the Church is established.

4   Acts 2:14-41.

5   John 20:19-23.

6   Acts 2.

7   Acts 5:17-42; Acts 12:1-19; Acts 16:16-40.

8   Acts 21:27-28:31.

9   The quote is derived from the document *Apologeticus* thought to have been written by Tertullian around 200 AD.

10   Romans 8:38-39.

11   Acts 1:11: 'Men of Galilee,' they said, 'why do you stand here looking into the sky? This same Jesus, who has been taken from you into heaven, will come back in the same way you have seen him go into heaven.'

12   Matthew 28:16-20.

13   James 5:14-16.

14   1 Thessalonians 5:16-18: 'Rejoice always, pray continually, give thanks in all circumstances; for this is God's will for you in Christ Jesus.'

15   Luke 24:46-48.

16   Matthew 28:18-20.

17   Philippians 4:6-7: 'Do not be anxious about anything, but in every situation, by prayer and petition, with thanksgiving, present your requests to God. And the peace of God, which transcends all understanding, will guard your hearts and your minds in Christ Jesus.'

18   The book of Revelation in particular testifies to this theme, as explored by Tom Wright in his commentary, *Revelation for Everyone*.

19   Proverbs 19:21: 'Many are the plans in a person's heart, but it is the Lord's purpose that prevails'.

20   2 Corinthians 12:9: 'But he said to me, "My grace is sufficient for you, for my power is made perfect in weakness." Therefore I will boast all the more gladly about my weaknesses, so that Christ's power may rest on me'.

21   Ephesians 5:18: 'Do not get drunk on wine, which leads to debauchery. Instead, be filled with the Spirit'.

22   1 Thessalonians 5:17: 'pray continually'.

23   Deuteronomy 8:3: 'He humbled you, causing you to hunger and then feeding you with manna, which neither you nor your ancestors had known, to teach you that man does not live on bread alone but on every word that comes

from the mouth of the LORD'; Matthew 4:4: 'Jesus answered, "It is written: 'Man shall not live on bread alone, but on every word that comes from the mouth of God."'

24   John 17:22-23; 1 Corinthians 12:12-14.

25   Hebrews 12:1-2: 'Therefore, since we are surrounded by such a great cloud of witnesses, let us throw off everything that hinders and the sin that so easily entangles. And let us run with perseverance the race marked out for us, fixing our eyes on Jesus, the pioneer and perfecter of faith. For the joy set before him he endured the cross, scorning its shame, and sat down at the right hand of the throne of God.'

26   Luke 9:23: 'Then he said to them all: "Whoever wants to be my disciple must deny themselves and take up their cross daily and follow me."'

27   https://www.alpha.org/ Accessed 9th Sept. 2021.

10

LIVING IN THE MOMENT

"What day is it?"
"It's today," squeaked Piglet.
"My favourite day," said Pooh.[1]

The biblical truths from the previous chapter were:

14. Jesus has been and will always be faithful to His promise to build His Church.

15. Living as a follower of Jesus involves being continually filled with the Holy Spirit, prayer, feeding on the Bible, being joined to our brothers and sisters within His body, the Church, and keeping our eyes fixed upon Jesus in order to follow His leading.

As we turn once more to our contemporary context, we begin in the driest desert of North America with a race for a million-dollar prize.

***

Anthony Levandowski first heard about the *DARPA* 'Grand Challenge' in 2003.[2] He was a graduate student in his early twenties at the time, studying engineering at the University of California, Berkeley. His innovative and entrepreneurial skills had already been in evidence for several years.[3] Whilst a teenager he had developed websites for local businesses. During his first year at university, he founded an internet services company which made fifty thousand dollars in its first twelve months. In his second year as an undergraduate he won first prize in a robotics competition by building a *Lego* robot which sorted *Monopoly* money. But it was the *DARPA* 'Grand Challenge' which would set him off on a pathway that would shape his next couple of decades.

DARPA is the US military research organisation whose funding produced Belle, the 'telekinetic monkey' we encountered in Chapter Four. Its 'Grand Challenge', as initially announced, was to produce a robotic vehicle which could race the entire 270 mile route from Los Angeles to Las Vegas for a million-dollar prize.[4] *DARPA* soon changed the race to a 100 mile, mainly off-road, section of the longer route within the Mohave Desert.[5] No vehicle would go on to claim the prize, and even the most successful of the entrants would complete less than eight miles of the course, yet the competition catalysed an explosion of interest in self-driving vehicles.[6]

While everyone else entered cars, trucks or buggies into the 2004 competition, Anthony Levandowski attempted to build a self-driving motorcycle.[7] Having forgotten to flip the switch to stabilise the system, Anthony watched in dismay as his creation fell to the ground at the start line preventing it from further involvement in the race.[8] Nonetheless he had already made a name for himself as the enterprising underdog with ambitious vision.

Following involvement in subsequent races, in 2009 Anthony Levandowski went on to co-found the *Google* self-driving car project which became *Waymo*, mentioned in Chapter Four.[9] Leaving *Google*

154

seven years later, Anthony went on to found another company which he subsequently sold to *Uber*. *Google* objected to the fact that before leaving, Anthony Levandowski downloaded thousands of files which then ended up in the hands of their rival. In 2020 Anthony pleaded guilty to one of the charges of which he was accused and was sentenced to eighteen months in prison, to be served once the COVID-19 pandemic had passed.[10] Before that happened, he received a full pardon from the outgoing US President in early 2021.[11]

As if to add further plotlines to a life which must surely be the subject of a biographical movie one day, in 2015 Anthony Levandowski created his own religion.[12] His assumption was that AI would soon surpass human abilities and that superintelligence was inevitable. The *Way of the Future* 'church' was for others who shared those beliefs and who, like him, sought to ensure a safe transition to the reign of 'super-intelligent' machines. He wanted those machines to know who "is friendly to their cause and who is not".[13] In other words this was a 'church', not only to further developments in AI, but also in which to worship AI and so make it less likely that it might go on to destroy humanity.

Mark Harris reported this conversation with the founder:

> "What is going to be created will effectively be a god," Levandowski tells me in his modest mid-century home on the outskirts of Berkeley, California. "It's not a god in the sense that it makes lightning or causes hurricanes. But if there is something a billion times smarter than the smartest human, what else are you going to call it?" ... "Humans are in charge of the planet because we are smarter than other animals and are able to build tools and apply rules," he tells me. "In the future, if something is much, much smarter, there's going to

155

be a transition as to who is actually in charge. What we want is the peaceful, serene transition of control of the planet from humans to whatever. And to ensure that the 'whatever' knows who helped it get along."[14]

At the end of 2020, the church was dissolved and its funds were given away.[15] *Way of the Future* had always been controversial with some people questioning whether its main aim was actually to distract from the legal battles surrounding Levandowski at the time. Nonetheless, the five-year existence of the church drew further attention not only to the issue of who or what we choose to worship, but also to the question of appropriate human engagement with AI.

***

The Bible describes the Church, the followers of Jesus, as the body of Christ, and assures us that there is just *one* body.[16] It speaks of unity. None of us is called to set up our own religious system. Neither are we expected to take responsibility for building what Jesus Himself has promised to establish. He calls all His followers to be involved in that work. He gives us well-fitting roles and He equips us to carry them out but He is the One who is responsible for bringing the growth.

I suspect that only very few of us are seriously considering founding our own religion. Many of us desire to see our local churches grow. Some of us are working to plant new congregations and to graft into existing ones. It is not unknown for new Christian denominations to form. Yet it is unusual to come across someone like Anthony Levandowski who sets out to establish a completely new religious enterprise. All of us, however, face the challenge of who or what we worship, and where we place our trust.

Anthony Levandowski helpfully draws our attention to important issues of Artificial Intelligence. We do need to reflect upon the direction of our technological growth. It may benefit us to consider in advance those questions and consequences which we may in time need to address. Yet we do not need to do so as if we do not know who to worship or who to trust.

Whatever AI developments may lie ahead, the Bible calls us to place our trust in God alone. He is the only One we are called to worship. Following Jesus at this point in history may involve working on issues of Artificial Intelligence but for each of us, no matter what our particular call, we are free to live as people who can trust that Jesus has already secured the future and He is faithful to His promise to build His Church.

***

Ray Kurzweil may not have advocated the creation of a new church but he has written of 'god' emerging from within creation through the universe 'waking up':

> Once we saturate the matter and energy in the universe with intelligence, it will "wake up", be conscious, and sublimely intelligent. That's about as close to God as I can imagine.[17]

This view portrays God as man's creation rather than the other way around. It describes humans continuing to learn and progress, with knowledge thus seeping into the universe itself such that 'god' is formed. A biblical poet warned against this arrogance when he wrote 'The fool says in his heart, "There is no God."'[18]

Whilst my own views differ fundamentally from those of Ray Kurzweil in this regard, I do believe that he offers us a very helpful lead in recognising that future technological developments are likely to bring about profound change. Central to his thinking is the significance of exponential growth. He uses the story of the Chinese Emperor to illustrate the power of this growth.[19]

The Emperor got so much pleasure from the new game of Chess that he wanted to reward the inventor. Unwittingly, the Emperor agreed to give the inventor a grain of rice on the first square of his chess board, two grains on the second square, four grains on the third square, and so on, doubling each time in an exponential pattern. To complete that pattern on the sixty-four squares of a chess board requires a staggering eighteen million trillion grains of rice, 18,000,000,000,000,000,000. That's enough rice to require paddy fields covering 'twice the surface area of the Earth, oceans included'.[20]

You may remember that when the human genome was first sequenced back in about the year 2000, initial progress was very slow. Many people expected it to finish behind target. Then suddenly progress sped up and the whole project was finished significantly ahead of time.[21] When the project reached the 1% completion point after seven years, there appeared to be an overwhelming amount of work still to do. Yet Ray Kurzweil predicted that because progress was doubling each year it would take just another seven years to finish.[22] He proved to be right because when technology moves ahead exponentially, it can take as long to complete the first 1% of the work as it can to finish the final 99%.[23]

When Kurzweil describes the world becoming unrecognisable by the mid-2040s we might be sceptical. We might point to *Give Vision* goggles, 'smart drugs' and microchip implants and suggest that current technology is still such a very long way from the future which

158

Kurzweil describes. Yet, if these current technologies represent just the first 1% of an emerging field, and if they act as an early indicator of the 99% of developments awaiting us in the future, then they may start to feel much more significant.

History demonstrates that many of us fail to take on board the implications of exponential growth. Even those professionals who are deeply embedded in technological fields can find it challenging to anticipate future developments. Quotes, like that attributed to the head of *IBM* in 1943 that there might be a worldwide market for five computers, have been taken out of context over the years.[24] Some have more reliable sources than others. Nonetheless, taken together, such statements do make a case for the fact that many technological predictions have proven wide of the mark.

Even in 1889, three years after the city of Rome had been electrified with alternating current (AC), Thomas Edison, an advocate of direct current (DC), was still quoted as declaring that 'fooling around with alternating current is a waste of time – no one will ever want to use it'.[25] In 1895, just eight years before Orville and Wilbur Wright made aviation history in the *Wright Flyer* near Kitty Hawk, North Carolina, Lord Kelvin is famously reported to have stated that 'heavier than air flying machines are impossible'.[26] In 1934, Einstein himself was quoted as saying that 'there is not the slightest indication that [nuclear energy] will ever be obtainable'.[27] In 1949, *Popular Mechanics* magazine published the opinion that computers in the future may 'perhaps weigh only 1.5 tons'.[28] In 1995, Robert Metcalfe, a holder of the US *National Medal of Technology*, not only predicted that the internet would collapse in 1996 but valiantly kept his promise to eat his words when his forecast was proved wrong.[29] More recently still, in just 2007 Steve Ballmer, former CEO of

Microsoft, expressed his belief that 'there's no chance that the *iPhone* is going to get any significant market share'.[30]

Whether our first reaction is to ridicule these experts who managed to get their predictions so wrong, or to be encouraged that we're not the only ones who make mistakes, the fact is that predicting the future of technology is difficult. Ray Kurzweil argues that one reason we find this so incredibly challenging is that, even though technological progress moves exponentially, we think linearly.[31] In other words, we think back to what has changed over the last ten, or twenty, or even a hundred years, and we assume that the change we'll see in the next ten, twenty or a hundred years will be similar. But Ray argues very convincingly that it will not, because technology does not develop in that way. Instead, it develops exponentially. Rather than technology moving forward by a set amount each year, it multiplies its reach during that time. Instead of moving ahead at a steady pace, it is always accelerating.

In 1965, Gordon Moore, co-founder of *Intel*, postulated that the number of transistors which can fit onto a silicon chip will double every year or so.[32] This observed pattern of exponential developments in computing power has become known as *Moore's Law*. The doubling time over the past decades has been a little longer than Gordon first estimated but the pattern has largely proven correct. It is why we hear that today's smartphones have more processing power than the multimillion-dollar supercomputers of just a few decades ago.[33]

Ray Kurzweil argues that the technological singularity which he envisions will come about if technological progress continues to accelerate in an exponential manner over the next few decades, as it has over the recent past. Will it? Is that realistic?

We know that technology can't keep on developing exponentially forever. Exponential growth occurs around us each day in different contexts but it never continues indefinitely. Bacteria can multiply exponentially for a period of their growth cycle and then their increase will slow in response to dwindling food resources. The same goes for exponential growth in other living and non-living systems.

We can be absolutely confident that technological growth will not follow an exponential pattern forever. So, is it unrealistic to expect rates of technological growth to continue for the next three decades? Not necessarily. The history of technology does indicate that very often as the rate of development of one technology slows down, another replaces it, so that overall, exponential development can continue for significant periods of time.[34]

For example, the technology of transport has developed since the eighteenth century as we have witnessed the emergence of first trains, then cars, followed by aircraft, rockets and spacecraft. In this way, the top speed of human-made vehicles has continued to increase exponentially over a two-hundred-year period.[35] In other words we cannot assume that the pace of technological change observed in the last few decades will end within the next few decades. Ray Kurzweil might be wrong. The exponential growth in computing we have experienced over the last six decades or so may not continue for another twenty-five years. But it is hard to argue with confidence that it will definitely slow within that time. If there is a possibility that such growth will continue, then Ray's view that the next two decades may bring profound technological change are surely worth considering.

***

How do we live with hope for tomorrow when we face such uncertainty? We simply focus on how Jesus has called us to live.

The first eight chapters of this book have already offered insights into what it means to live as followers of Jesus in our contemporary world. We are to live in the light of the biblical truths of Creation, God's Law, Jesus' incarnation, death and resurrection, and of New Creation.

If we follow Jesus in the light of Creation, we celebrate the goodness of the world and its life, we recognise the brokenness which has come about through human selfishness and disobedience, and we face the reality of our human role to represent God as creatures made in His own image. Following Jesus involves caring for His Creation.

If we follow Jesus in the light of God's law, we obey the commands to love God with all our heart, mind, soul and strength, and to love our neighbour as ourselves. We honour the goodness of the way in which God has created humans and the world, and we act for the good of others as well as ourselves.

If we follow Jesus in the light of His incarnation, we have further reason to honour and respect the human form. In our dealings with one another and with machines, we cherish the precious worth of each and every human being. We'll seek to be fully human and fully alive knowing that He models what it means to do that. Following Jesus will involve being prepared to live differently. It may involve being willing for us and even for our children to remain as we are if others move ahead to seek to become more than human.

If we follow Jesus in the light of His death and resurrection, and the New Creation they bring about, we live in certain hope of the future which Jesus has already secured. We are liberated to live as He has called us to live, without any unrealistic pressure to bring about the ultimate future through our own actions. We do not need to carry the weight of the future on our own shoulders.

So too, if we follow Jesus in the light of the fact that this is the time of the Church, we focus on the freedom of simply being obedient to His call which involves participating in what God is doing. Jesus is at work in and through His Church and each of us is invited to be part of this body.

The second of the two biblical truths highlighted in the previous chapter was that living as a follower of Jesus, part of the Church, involves being continually filled with the Holy Spirit, prayer, feeding on the Bible, being joined to our brothers and sisters within His body, and keeping our eyes fixed upon Jesus in order to follow His leading. Those are challenging demands. If we attempt to meet them, we're likely to fall down at times and to need God to pick us up and to place us on our feet once more. Being obedient to such a call is the work of a lifetime. Yet such a call is the well-fitting yoke which Jesus promised.[36]

The Bible is a gift which helps enable us to respond to that call. The fifteen biblical truths which have been particularly emphasised within this book are listed together within Appendix One. Asking the Holy Spirit for His guidance, that list may be helpful when reflecting upon responding to God's call in our contemporary context.

Such a call does not include needing to predict the future with detailed accuracy. Neither does it include carrying the weight of technological concerns on our own shoulders. Jesus invites us to know the freedom of trusting Him with the burdens of the future, and simply following.

Doing so will mean that we engage with technology in different ways as we draw upon the unique combination of gifts that each of us has received. Some of us will be called to a particular focus on lobbying governments and influencing our legal frameworks. Others of us will have key roles, as parents, grandparents or teachers, in nurturing and educating our children and in equipping them for fullness of life, now and in the future. Some of us will be involved in leading churches. Others will be working as scientists, engineers, nurses, doctors and related roles. As is explored is Appendix Three, those working professionally on these various technological frontlines will gain particular insights which may help the rest of us to make sense of our technological future. Similarly, we in turn may be able to offer those professionals support as they work on our behalf.

Whatever our work, all of us will face some sort of technological engagement. Living as a follower of Jesus involves living in the world.[37] Our lives are shaped and affected by our technological context. It's right that we engage with that appropriately.

We do not need to be overwhelmed by the technological concerns of today and tomorrow yet neither is it right to shut our eyes. If the wider world is to be able to engage with technological challenges drawing upon biblical wisdom and the insights of the Church, then Jesus' followers need to be engaging with these issues and speaking into them. Even if the world is not interested in our insights, we are still responsible for being salt and light[38] and so modelling what it means to follow Jesus at this time in history.

As it can be uncomfortable viewing the news, so too it can be unsettling to grapple with the realities of our technological world. The main reason for drawing attention to both the concerns and the promise of our technological context is that we may see clearly the realities of the world in which we are called to live. Following Jesus involves opening our eyes to others and to His creation, trusting Him and being obedient, not in some imagined existence, but in the reality of our day-to-day interconnected lives.

We need to walk a tightrope of technological engagement, avoiding falling off on one side by so failing to engage that we miss the changing nature of our contemporary world, whilst also making sure we don't drop on the other side by being so overwhelmed by technological issues that we forget that this is the time of the Church. It is not healthy to spend all of our time glued to the news yet if we're to express our solidarity with those around the world living with the consequences of poverty, war, persecution or natural disaster it helps to know something of those realities. If we're to live well in a world increasingly shaped by technology, we may need to engage more deeply with that context too. Doing so from the perspective that this is the time of the Church can enable us to respond with a perspective of hope and of freedom and of partnership.

Repeatedly the Bible encourages us not to fear. Though we already face and will continue to face personal and global challenges we also know the end of the story. We know that Jesus has already secured the ultimate victory through His death and resurrection. So we face the future with hope.

What it means to be the Church at this point in history will emerge as each of us who seek to follow Jesus respond with obedience to His call, fully alive in the world of today and tomorrow. I pray that this book might be useful to all who are seeking to do that.

God bless you.

Justin, September 2021.

SUMMARY

1. Anthony Levandowski, a key figure in the world of autonomous vehicles, once set up his own 'church' as part of his own engagement with technology.

2. Ray Kurzweil has written of the universe 'waking up' and so of 'god' emerging from within creation.

3. Exponential growth can bring about change at a pace that our brains can find hard to grasp.

4. The future of technology is notoriously difficult to predict.

5. Those working professionally in the fields of science, medicine and technology will have insights into these areas.

6. What it means to live in the time of the Church is deeply shaped by our contemporary contexts.

7. The Bible is a gift which helps us to follow Jesus at this point in history.

8. We are called to live with hope not fear.

QUESTIONS FOR REFLECTION

1. What question would you like to put to Anthony Levandowski about the *Way of the Future* 'church'?

2. Do you think a super-intelligent machine would like to be worshipped? Why?

3. What helps you to live with hope rather than fear?

4. What do you need to reflect upon further in the light of this book?

5. Are there any actions you feel prompted to take at this point?

1 The source of this quote attributed to these A. A. Milne characters appears to be unclear.
2 Alex Davies, *Driven*, p6.
3 Ibid., p27-29.
4 Ibid., p9-23.
5 Ibid., p23-26.
6 Ibid., p59-77.
7 Ibid., p45-52.
8 Ibid., p76-77.
9 Ibid., p133-155.
10 Ibid., p246-248.
11 https://www.bbc.co.uk/news/technology-55732450 Accessed 9th Sept. 2021.
12 https://www.wired.com/story/anthony-levandowski-artificial-intelligence-religion/ Accessed 9th Sept. 2021.
13 The website of the *Way of the Future* church, from which this quote was obtained, is no longer accessible at the time of writing.
14 https://www.wired.com/story/anthony-levandowski-artificial-intelligence-religion/ Accessed 9th Sept. 2021.
15 https://techcrunch.com/2021/02/18/anthony-levandowski-closes-his-church-of-ai/ Accessed 9th Sept. 2021.
16 Romans 12:4-5; 1 Corinthians 12:12-31; Ephesians 4:1-16; Colossians 1:18.
17 Ray Kurzweil, *The Singularity is Near*, p375.
18 Psalm 14:1a.
19 Ray Kurzweil, *The Age of Spiritual Machines*, p36-39.
20 Ibid., p36.
21 http://news.bbc.co.uk/1/hi/sci/tech/2940601.stm Accessed 9th Sept. 2021.
22 https://www.kurzweilai.net/waking-the-universe Accessed 9th Sept. 2021.
23 Francis Collins is the scientist who led much of the work in his role as director of the National Human Genome Research Institute. He has written about his Christian faith in the 2007 book *The Language of God* and of personalized medicine in the 2010 publication *The Language of Life*. He was interviewed by Nicky Gumbel within the 2021 *Leadership Conference* https://www.alpha.org/blog/leadership-conversations-with-nicky-gumbel-podcast-francis-collins/ Accessed 9th Sept. 2021.
24 https://geekhistory.com/content/urban-legend-i-think-there-world-market-maybe-five-computers Accessed 9th Sept. 2021.
25 https://libquotes.com/thomas-edison/quote/lbx5e7q Accessed 9th Sept. 2021.
26 https://www.nasa.gov/audience/formedia/speeches/fg_kitty_hawk_12.17.03.html Accessed 9th Sept. 2021.
27 https://www.nytimes.com/1964/08/02/archives/the-einstein-letter-that-started-it-all-a-message-to-president.html Accessed 9th Sept. 2021.

28 https://www.popularmechanics.com/technology/a8562/inside-the-future-how-popmech-predicted-the-next-110-years-14831802/ Accessed 9th Sept. 2021.

29 https://1995blog.com/2015/12/03/prediction-of-the-year-1995-internet-will-soon-go-spectacularly-supernova/ Accessed 9th Sept. 2021.

30 https://usatoday30.usatoday.com/money/companies/management/2007-04-29-ballmer-ceo-forum-usat_N.htm Accessed 9th Sept. 2021.

31 Ray Kurzweil, *The Singularity is Near*, particularly Chapters 1 and 2.

32 https://www.britannica.com/biography/Gordon-Moore Accessed 9th Sept. 2021.

33 https://www.webfx.com/blog/internet/how-much-did-the-stuff-on-your-smartphone-cost-30-years-ago/ Accessed 9th Sept. 2021.

34 Ray Kurzweil, *The Singularity is Near*, p43-46.

35 http://www.foresightguide.com/superexponential-growth-j-curves/ Accessed 9th Sept. 2021.

36 Matthew 11:28-30.

37 John 17:13-19.

38 Matthew 5:13-16.

EPILOGUE: A SERMON

This is a sermon I gave at Lee Abbey, Devon, on Thursday 7th March 2019 towards the end of the house-party which helped catalyse the writing of this book.

What are you hoping for? In the short term your hopes may simply be that I won't speak for too long! But what are your hopes and fears for beyond that, for what comes tomorrow and the next day and further ahead? For those of us here as guests that will involve looking ahead to tomorrow or whenever we head off. For those of you on community it involves looking beyond this house-party.

What are your hopes and fears for the future, for yourself, for those you love, for our home countries, for our world?

Have you followed the news story concerning the comic series *Second Coming*? Over the last few months news emerged that *DC Comics*, famous for their depictions of superheroes such as Superman and Wonder Woman, was preparing a new series, called *Second Coming*. In this series a new superhero, Sun-Man, would team up with Jesus to try to save the world. Seriously! The idea for the story is that God was upset with Jesus's performance the first time he came to Earth, since Jesus was arrested so soon and crucified shortly after. Seeing Sun-Man's performance, the Father thinks Jesus could learn a thing or two and sends him back to Earth for another try.

Understandably, a number of people were upset by Jesus being depicted in that way, and exactly two weeks ago it was announced that, because of the effect of a large online petition, *DC Comics* are not going to publish the story. The writer is still looking to get it published elsewhere but *DC Comics* won't be involved.

Different Christians have different reactions to the idea of a story like this being published. There were some who signed the petition who felt the story was blasphemous and so needed to be stopped. Others published comments online raising the possibility of using a story like this to talk to work colleagues and others about the true gospel story, to engage with what the world is saying about Jesus and to use that as a springboard for talking about faith.

Whatever we think about the fact that a story like this has been written, I think it poses two questions for all of us. Firstly, how is it that after two thousand years of the Church speaking out the good news of Jesus, the cross and the resurrection can be looked upon as a failure? The Bible gives us an answer to that first question.

As Paul writes in 1 Corinthians 1:18, 'For the message of the cross is foolishness to those who are perishing, but to us who are being saved it is the power of God'. As Jesus Himself said in relation to parables, in Matthew 13, quoting from Isaiah: 'This is why I speak to them in parables:

> "Though seeing, they do not see; though hearing, they do not hear or understand." In them is fulfilled the prophecy of Isaiah:
>
> "You will be ever hearing but never understanding; you will be ever seeing but never perceiving. For this people's heart has become calloused; they hardly hear with their ears, and they have closed their eyes. Otherwise they might see with their

eyes, hear with their ears, understand with their hearts and turn, and I would heal them."

But the story also poses another question, a more personal one. As we look ahead into the future does the power of the cross and the resurrection liberate us to look ahead with hope and to live with hope? If we declare with our lips that through the cross and the resurrection Jesus has already achieved the ultimate victory, that we're free from sin and death, that Jesus has already secured the end of the story and that it involves a new earth and a new heaven, does that reality set us free to live in hope?

That question is important. It's important for us and for those around us. And it's important for the world. As those of us who have been meeting together over these last few days have been exploring, we're living in a world of incredibly rapid technological change, of environmental concerns. That's not to mention political uncertainties and all the personal and family anxieties we may be carrying.

Are we living as people of hope? Is hope infusing our own circumstances, and are we living as people of hope, shining out the light of hope into the world as a gift for others?

The reality is that Jesus both enables and empowers us to live that way. He has set us free. Far from being a failure, the cross and the resurrection are ultimate good news and they speak exactly into all that we face today and tomorrow.

As some of us explored this morning, as machines become able to carry out more and more tasks which were once something that only people were able to do, tasks like play chess, or drive a car, write a poem, we face another question of our own. Do we one at a time, relinquish those activities and say that as humans we'll leave those to the machines, or do we recognise that our challenge is to seek to be more fully human ourselves? As Christians we worship the One who is most fully human of all. The One who was described by Paul in his letter to the Philippians in these words:

> ... being in very nature God, did not consider equality with God something to be used to his own advantage; rather, he made himself nothing by taking the very nature of a servant, being made in human likeness. And being found in appearance as a man, he humbled himself by becoming obedient to death — even death on a cross! Therefore God exalted him to the highest place and gave him the name that is above every name, that at the name of Jesus every knee should bow, in heaven and on earth and under the earth, and every tongue acknowledge that Jesus Christ is Lord, to the glory of God the Father.

We worship the One who is fully human and who calls us to be fully alive. We're called to love God with all our heart, mind, soul and strength and to value His creation and our own nature as a creature made by a master craftsman. We're called to love our neighbour as ourselves. And we're called to look ahead with hope, because Jesus has already achieved the final victory and has already opened up the doorway to a new heaven and a new earth.

I think it involves remembering that as we love God with heart, mind, soul and strength, we're recognising that He is the perfect Creator and He made us. He looked on us and on creation and pronounced that creation, us included, as very good. I think it involves recognising that we're social creatures, created for relationship with God and with others. We're called to love our neighbours as ourselves. And we're called to live in the light of resurrection hope and the open doorway to a new heaven and a new earth.

We live in a changing world but we worship Emmanuel, God with us, who knows what it is to be fully human and what it is to be fully alive. It's He who gives us hope for tomorrow.

How can we do that as we go out from this place to live in hope? I want to offer four ideas.

Firstly, take something from here! And I don't mean the towels or the cutlery! When Delana and I were here on community, final notices at the last meal of a house-party would involve the request to take what's yours and to leave what's ours! I hope that all of us who are leaving have a sense of going out with more than we came with. What new insights has God given you this week? Treasure that and hold onto it. Let it be a gift in what comes next.

Secondly, go out delighting in the way God has made you. He's called us to be fully human and fully alive. He himself models what it means to be fully human. In a world in which machines can mimic humans, the emphasis is on us to be more fully human, not less. And we are each of us fearfully and wonderfully made. Do you know of Simon Guillebaud who has been working for many years with the church in Burundi? Some months back now Simon reported of a successful cow project in Burundi. Long-horned cows are a symbol of wealth in that nation but they produce little milk and they overgraze the land. A Christian called Evariste sensed God saying that "the problem is cows and the answer is cows". Despite setbacks, Evariste persuaded his village to raise Friesian cows, of less symbolic value, but producing more milk and resulting in less overgrazing. As a result, the village has been transformed and it's now thriving. The villagers declare that as

God once spoke through a donkey, so now He speaks to them each day through cows. If we look, we'll see God's faithfulness and His promise revealed in the details of creation. And we'll see His goodness revealed in each of us too. We are fearfully and wonderfully made by a master craftsman. His strength is made perfect in weakness.

Thirdly, be patient and peaceful in the waiting. Both with yourself and with God. His timing is good. Do you know that the Great Wall of China was built over a period of more than two thousand years? Anna Watkin of the *London Institute of Contemporary Christianity* wonders what it would have felt like to be a stone mason in the middle of that project, looking into the far distance, looking at the small stone in one's own hand. Wondering how many more stones would be needed on this section alone, thinking about the long decades when it felt as if the task would never be finished. Whatever we feel as we look at our own lives and at the world around us, God is working His good plans and purposes out. His timing is good. He has already achieved the final victory. Be patient and peaceful in the waiting. Lay the stones which He calls you to lay as part of what He's bringing about but know that the project management of the whole is on His shoulders, not ours! He's promised to build His church and He is faithful in doing that.

Fourthly, be open to being surprised by God in relation to what He can do through you. I came across the incredible true story of two identical twins, two German brothers, Matthew and Michael Youlden, who have got a real gift for learning other languages. They live in Berlin where there's a large Turkish population so they set themselves the challenge of learning to speak Turkish in just one week! Their aim was to be able to have a full conversation with someone whose first language is Turkish just one week from starting their challenge. At the end of the week, they met with a presenter from Turkish television and were quizzed by him, in Turkish, on Turkish pop culture. They did it. You can watch both some of that conversation and their work towards it online. It may not be through foreign languages, I'm sure it won't be for me, but for each of us God can do incredible things in and through us if we make ourselves available to Him.

So, let's go out from here in hope, because of who He is and because of the power of the cross and the resurrection. Let's hold onto all that He's whispered to us this week. Let's go remembering that we are fearfully and wonderfully made, confident in His timing, and knowing that as we make ourselves available to Him, He can do incredible things in and through us.

Let's pray.

APPENDIX 1

The biblical truths specifically identified within this book

1. Creation is good and Jesus is our Creator God in human form.

2. Humans are creatures.

3. Creation is marred as a consequence of the fall.

4. Humans are made in the image of God and so have a role as representatives of God.

5. We are commanded to love God with all our heart, mind, soul and strength.

6. We are commanded to love our neighbour as ourselves.

7. In Jesus, God Himself became a human being.

8. Jesus grew at the pace of human childhood.

9. Jesus' earthly life was a life lived in community.

10. Jesus calls us to be fully alive as human beings and He models for us what it means to be so.

11. Out of love for each of us, Jesus died and then rose again, not avoiding death but going through it.

12. Jesus' body was transformed, after His resurrection from the dead, into His 'resurrection body'.

13. Jesus' death and resurrection have already secured the future.

14. Jesus has been and will always be faithful to His promise to build His Church.

15. Living as a follower of Jesus involves being continually filled with the Holy Spirit, prayer, feeding on the Bible, being joined to our brothers and sisters within His body, the Church, and keeping our eyes fixed upon Jesus in order to follow His leading.

APPENDIX 2

Technology, the Environment and the overarching story of the Bible

Whilst the focus of this book is fixed firmly upon the subject of technology, I recognise that it may prompt environmental questions and reflections, not least through its exploration concerning creation in Chapters One and Two. This appendix is intended to offer a few signposts for further engagement with the subject.

***

What might you do if you had tens of billions of dollars available to you, and some time on your hands?

Alongside his philanthropic work around the globe, drawing on resources generated by his success in the world of technology, Bill Gates decided to focus a significant part of his attention on the environment. A decade of investigating the causes and effects of climate change led, first, to the launch of the company *Breakthrough Energy*, an 'effort to commercialise clean energy and other climate-related technologies' and then, six years later, to the publication of his 2021 book, *How to avoid Climate Disaster*. Bill describes the challenge as he sees it:

There are two numbers you need to know about climate change. The first is 51 billion. The other is zero.

Fifty-one billion is how many tons of greenhouse gases the world typically adds to the atmosphere every year. Although the figure may go up or down a bit from year to year, it's generally increasing. This is *where we are today*.

Zero *is what we need to aim for*. To stop the warming and avoid the worst effects of climate change – and these effects will be very bad – humans need to stop adding greenhouse gases to the atmosphere.[1]

Bill summarises a position which is held by increasing numbers of leaders around the world. During recent years more and more people are coming to see climate change as one of the key challenges facing the world today. Where Bill Gates parts company with some of those others is in being hopeful of overcoming that challenge without turning our backs on an energy-rich lifestyle:

Although heavy hitters like me should use less energy, the world overall should be using *more* of the goods and services that energy provides. There is nothing wrong with using more energy as long as it's carbon-free. They key to addressing climate change is to make clean energy just as cheap and reliable as what we get from fossil fuels. I'm putting a lot of effort into what I think will get us to that point and make a meaningful difference in going from 51 billion tons a year to zero.[2]

He continues:

Most experts agree that as we electrify other carbon-intensive processes like making steel and running cars, the world's electricity supply will need to double or even triple by 2050. And that doesn't even account for population growth, or the fact that people will get richer and use more electricity. So the world will need much more than three times the electricity we generate now.[3]

Bill Gates does acknowledge a role for living more simply, but it's far from being his key focus.[4] He has come to acknowledge the value of reducing the demand for energy wherever we can yet he still stresses the importance of 'progress', despite its associated increase in energy consumption. He describes the rapid growth of Shanghai and then argues that:

This story is being repeated all over the world, though the growth in most places isn't as dramatic as it was in Shanghai. To repeat a theme which comes up repeatedly in this book: *This progress is a good thing* ... people's lives are improving in countless ways. They are earning more money, are getting a better education, and are less likely to die young. Anyone who cares about fighting poverty should see it as good news.[5]

Yet not everyone agrees. There isn't consensus about whether such growth is good news, or whether it is possible to care for the environment without reducing global consumption. Another Bill sees things differently.

We encountered Bill McKibben in Chapter Four in the context of questioning whether or not we can say 'Enough' to technology. We drew upon his examples of China and ship building in the fifteenth to twentieth centuries, of Japan and guns in the seventeenth to twentieth centuries, as well as that of the Amish during the last hundred years or so, and of global engagement with nuclear weapons since the Second World War.

Bill McKibben is a writer for whom the environment has been a key theme throughout his career. He is the founder of environmental organisations *Step it Up* and *350.org*. The latter of those two organisations is named after a target level for carbon dioxide in the earth's atmosphere. Bill McKibben describes how, in late 2007, James Hansen, a leading climatologist named 350 parts per million of carbon dioxide as a safe upper limit.[6] Global figures had actually increased to above that threshold way back in April 1988 and by December 2007, when James Hansen made his statement, it stood at 385 parts per million.[7] At the time of writing this book, the latest figure from June 2021 is of 416 parts per million.[8]

When he heard that statement, Bill McKibben concluded that the planet was already so far beyond James Hansen's safe limit that it was right to face the reality that we had already passed beyond a point of no return. He writes of recognising that: 'we'd never again inhabit the planet I'd been born on or anything close to it'.[9] He went on to propose renaming our planet as *Eaarth*, with two 'a's, to acknowledge the different nature of this new home.[10] He believes that it may be possible, 'if we're very lucky and very committed', to reduce our carbon dioxide levels back down to levels last seen in the late 1980s.[11] Whether or not we achieve that, he argues, our planet has already been irreversibly changed.

So, what's Bill McKibben's proposal? How does he suggest we live in this new world? Unlike Bill Gates, he calls for lifestyles which are 'small'.[12] He describes our contemporary culture as enamoured with 'fast', 'big' and 'growth'. He calls us instead to live differently:

184

We lack the vocabulary and the metaphors we need for life on a different scale. We're so used to *growth* that we can't imagine alternatives; at best we embrace the squishy *sustainable*, with its implied claim that we can keep on as before. So here are my candidates for words that may help us think usefully about the future.

Durable
Sturdy
Stable
Hardy
Robust

These are squat, solid, stout words. They conjure a world where we no longer grow by leaps and bounds, but where we *hunker down* …

… the economy that has defined our Western world is like a racehorse, fleet and showy. It's bred for speed … But don't put it on a track where the rain has turned things muddy; know that even a small bump in its path will break its stride and quite likely snap that thin and speedy leg … What we need to do, even while we're in the saddle, is transform our racehorse into a workhorse – into something dependable, even-tempered, long-lasting, uncomplaining. Won't go fast, will go long; won't win the laurel, will carry the day … We're talking walk or trot or jog, not canter or gallop.[13]

Bill McKibben writes of living 'lightly, carefully, gracefully'[14] and he summarises how he sees the challenge we face:

My point throughout this book has been that we'll need to change to cope with the new Eaarth we've created. We'll need, chief among all things, to get smaller and less centralized, to focus not on growth but on maintenance, on a controlled decline from the perilous heights to which we've climbed.[15]

Two Bills. Both drawing our attention to the environment as a key challenge for our world today. Each offering very different visions of how we overcome that challenge. One seeing the further pursuit of technological and economic growth as a key part of the solution, the other proposing that its only by forgoing further growth and focusing on living more lightly, that we'll be able to move ahead safely.

***

How do we engage with these visions with discernment? Are there aspects of each which we might embrace?

It would take another book to explore more fully how the overarching story of the Bible might help us to engage with the environment, but a rough sketch may be enough to point towards such an exploration. That sketch might involve the following sorts of outlines and broad-brush strokes:

- The biblical theme of Creation and the Fall points to an awesome world, which whilst subject to threat and decay, is of immeasurable value.[16]

- The biblical theme of the Law identifies human disobedience and points to the reality that we have not done all we needed to, to care for the planet.[17] We live with the consequences of that sobering reality.

- The biblical theme of Incarnation points to the immense worth of each and every human being. Our care for creation needs to be informed by the fact that our impact upon the lives of people today and in future generations matters.[18]

- The biblical theme of Jesus' resurrection and the New Creation points to the fact that this world has eternal worth. God brings New Creation out of the old, not by throwing away what has gone before, but by transformation of what we have now.[19] Jesus restores what is broken but that does not take away the importance of how we treat the world.

- The biblical theme of the Church points to the fact that we have work to do today, even as we trust God to bring about the future. As well as repenting for our past actions we are called to care for the world well from this point in time.[20]

It seems to me that as soon as we begin to engage with the environment from a biblical perspective, we very quickly face the imperative to care for creation. The Bible calls us to rise to the task of being good stewards of life on earth.[21] It describes a core human calling to work for the good of creation. That creation includes people, animals, plants, rainforests, mountain ranges, arctic tundra, and ocean depths as well as the trees on our streets and the air in our cities.

We're called to care for creation because it doesn't belong to us! As a biblical poet reminds us, 'The earth is the LORD's, and everything in it'.[22] It's lent to us in order that we have a home. It's ours not to possess but to steward and to pass on to future generations in at least as good a condition as that in which we received it, hopefully in a better one.

Caring for all aspects of creation is a core biblical principle of right human engagement with the environment. It strikes me that there is an interesting difference between a biblical engagement with technology and the environment in this respect. At the very start of our biblical engagement with the environment we confront the need to care for creation as stewards. This biblical call seems to resonate with a common human appreciation of the natural world. I imagine that most of us, regardless of our spiritual, religious or theological beliefs, are willing to agree that clean air and clean water are of immense value. We appreciate the diversity of plant and animal species. Where we may disagree is on questions of how to care for the planet, what sacrifices we are prepared to make to do so, and who needs to cover those costs.

With technology, I believe we start from a place of less consensus. Is there a shared position on technology which is comparable to a common belief in the value of the environment? I'm not sure there is. Nonetheless, for whoever wishes to draw upon its wisdom, the overarching story of the Bible offers a shared foundation from which to explore technology.

That same biblical foundation offers a means to develop our environmental thinking too. And we need that because whilst our environmental discernment may begin from a shared appreciation of the natural world, as we know, we quickly find ourselves in disagreement. As we have already started to explore, even those who agree on the significance of addressing climate change, can disagree on how to go about doing that.

How might the overarching story of the Bible sketched out above help us to engage with the differing views of Bill Gates and Bill McKibben on caring for the environment?

188

Seeing creation as being of immense value is likely to make us receptive to the call of each of them to address environmental concerns. We may disagree on details but that biblical perspective of the goodness of creation will lead us to affirm their commitment to care for our natural world.

The biblical theme of the law may illuminate the question of whether or not environmental change can be brought about without reducing our energy consumption. Setting out to love God with all our heart, mind, soul and strength may help us live in such a way that our energy use is pleasing to Him rather than an expression of greed or an unhelpfully lavish lifestyle. Setting out to love our neighbour as ourselves may help us to reflect on how our energy use relates to that which is available for others. Specific laws such as that to avoid harvesting the edges of our fields,[23] may help point us towards a lifestyle not of maximum efficiency, but rather of generosity and community.

The theme of the Incarnation may draw our attention to the value of each and every human being, all those alive today, no matter where they live, as well as future generations. This focus might prompt us to explore who benefits from continued economic growth, both now and in years to come. It might move our attention away from growth as such, to the beneficiaries of such growth. Such a focus might also prompt us to question whether a move towards maintenance and hunkering down is consistent with lifting others out of poverty. If affluent nations and individuals committed to living more lightly, what impact would that have upon those less fortunate, and on future generations?

The biblical themes of crucifixion, resurrection and new creation might draw attention to uses of energy which reflect placing our hope

189

for the future in places other than God's hands. For example, whilst noting that motivations for any projects will be widely diverse, we might nonetheless question the energy and other resources spent on cryogenics, or cryptocurrencies, wherever such projects are based upon a desire to take the future into our own hands.

The biblical theme of the Church might help us to reflect upon how our energy use reflects our call and priorities as followers of Jesus at this point in history. Does that use express a focus upon, or a distraction from, prayer, service, nurturing community, feeding on the Bible, *etc*?

I am aware that the outline engagement with these biblical themes in this appendix raises more questions than it answers, but I hope to have demonstrated that the overarching story of the Bible can illuminate not just technology, but any issue we may need to reflect upon. Furthermore, the biblical story not only calls us to care for creation but offers illumination for engaging with specific questions such as how we use energy and other resources. To further illustrate this point, let's consider the contentious subject of geoengineering.

Geoengineering is the use of global-scale technological solutions to environmental issues, for example seeking to cause rainfall by triggering water droplets to form in the atmosphere, mimicking the effects of volcanic activity so as to block out sunlight, fertilising the oceans in order to capture carbon, *etc*. Geoengineering represents an area of common ground between the technological focus of this book and the environmental subject of this appendix. It is held up by certain technologists as a solution for environmental concerns. Others describe it as 'hugely high risk and exactly the kind of reckless, short-term thinking that got us into this mess'.[24]

The overarching story of the Bible might prompt us to ask the following sorts of questions: How might such a project relate to love of God and a respect for His creative ability? Does this project express love for neighbours, both those alive today, and future generations? Is this project an expression of trust in God to bring about the future, or an attempt to take that into our own hands? How does this project relate to the call of the Church at this point in history?

Reflecting on God's creative ability and trusting Him to bring about the future are likely to caution against rushing too quickly into any projects which might risk damaging our world yet further.

***

As well as the previously-mentioned writing of Bill Gates and Bill McKibben in this area, there are, of course, so many other sources of environmental stories which might provide stimulating insights for a further engagement with this subject.

John Houghton, who died at the age of eighty-eight in April 2020, was an atmospheric physicist who co-chaired the Intergovernmental Panel on Climate Change's scientific assessment working group which shared the Nobel Peace Prize in 2007. He was also president of the John Ray Institute, connecting environment, science and Christianity. His book entitled *Global Warming* is still a key publication in the field.[25] Alastair McIntosh also explores 'the science of what is happening to the planet today' within his 2009 book, *Hell and High Water*.[26]

Paula Clifford looks at the subject through the lens of the messages to the seven churches in the Book of Revelation in her 2009 book, *Angels with Trumpets*.[27] New Testament scholar Richard Bauckham explores how to 'read the Bible in an age of ecological disaster' within his 2010 book, *Bible and Ecology*.[28]

Nick Spencer and Robert (Bob) White have worked together to write *Christianity, Climate Change and Sustainable Living*, a 2007 publication exploring sustainable consumption and production.[29] Nick Spencer is currently Senior Fellow of the think tank *Theos*, and connected to the *Jubilee Centre* and the *London Institute of Contemporary Christianity*. Professor Bob White is a Professor of Geophysics and Emeritus Director of the *Faraday Institute for Science and Religion*. Bob White has also partnered with Jonathan Moo, Professor of Theology, to write the 2013 publication, *Hope in an Age of Despair* exploring 'the gospel and the future of life on earth'.[30]

George Marshall explores issues of 'why our brains are wired to ignore climate change' within his 2014 book, *Don't Even Think About It*.[31] Naomi Klein explores issues of capitalism and the climate in *This Changes Everything*, also published in 2014.[32]

Pope Francis' 2015 encyclical *Laudato Si* is a significant milestone in writing on Christian care for creation, and triggered action not only within the Roman Catholic community but throughout the wider church.[33] The 'five marks of Anglican mission' includes the call 'to strive to safeguard the integrity of creation, and sustain and renew the life of the earth'.[34] These marks continue to prompt action and reflection on the subject.

A Rocha is one of a number of organisations seeking to put into practice the core Christian belief that our world matters, through projects such as Eco-Church.[35]

Key figures within our society have helped stimulate engagement with these areas. David Attenborough's advocacy for creation care has led to the emergence of the phrase the 'Attenborough Effect' to describe his impact upon plastic use in particular.[36] Greta Thunberg has had a similarly profound influence upon society, mobilising both young and old to take action for environmental care.[37]

These and other sources of stories on environmental care might be used to stimulate further reflection on the subject through engagement with the biblical overview described throughout this book and within this appendix.

1   Bill Gates, *How to avoid a Climate Disaster*, p3.

2   Ibid., p15.

3   Ibid., p79.

4   Ibid., p95-96.

5   Ibid., p102.

6   Bill McKibben, *Eaarth*, p15.

7   https://www.climate.gov/news-features/understanding-climate/climate-change-atmospheric-carbon-dioxide Accessed 9th Sept. 2021.

8   https://www.climate.gov/news-features/understanding-climate/climate-change-atmospheric-carbon-dioxide Accessed 9th Sept. 2021.

9   Bill McKibben, *Eaarth*, p15.

10   Ibid., p2.

11   Ibid., p16.

12   Ibid., p102-150.

13   Ibid., p102-103.

14   Ibid., p151-212.

15   Ibid., p204.

16   Genesis 1-3.

17   Passages such as Isaiah 24:4-6 express the biblical link between whether or not we follow God's ways and the impact of those actions on the health of the earth: 'The earth dries up and withers, the world languishes and withers, the heavens languish with the earth. The earth is defiled by its people; they have disobeyed the laws, violated the statutes and broken the everlasting covenant. Therefore a curse consumes the earth; its people must bear their guilt. Therefore earth's inhabitants are burned up, and very few are left'.

18   Ezekiel 34:18-19: 'Is it not enough for you to feed on the good pasture? Must you also trample the rest of your pasture with your feet? Is it not enough for you to drink clear water? Must you also muddy the rest with your feet? Must my flock feed on what you have trampled and drink what you have muddied with your feet?'

19   In *Bible and Ecology*, p175, the theologian Richard Bauckham writes of the Apostle Paul seeing New Creation as being brought about by transformation rather than replacement in these words: 'Paul plainly does not regard new creation as the replacement of the present creation by a quite different, new one. If he did, he would have to suppose that in Christian conversion one human being is replaced by a brand new one. He evidently sees new creation as the eschatological renewal of creation. The vivid language of old things passing away and all things becoming new refers to a transfiguration of reality into a new form. It is radical transformation here, but not replacement'.

20   Passages like Romans 3:31 emphasise the ongoing call to obey the law: 'Do we, then, nullify the law by this faith? Not at all! Rather, we uphold the law'. The law includes commands such as this one from Leviticus 26:3-4: 'If you follow my

decrees and are careful to obey my commands, I will send you rain in its season, and the ground will yield its crops and the trees their fruit'.

[21] Genesis 1:26-29.
[22] Psalm 24:1a.
[23] Leviticus 23:22: 'When you reap the harvest of your land, do not reap to the very edges of your field or gather the gleanings of your harvest. Leave them for the poor and for the foreigner residing among you. I am the LORD your God.'
[24] Naomi Klein, *This Changes Everything*, p58.
[25] John Houghton, *Global Warming*.
[26] Alastair McIntosh, *Hell and High Water*.
[27] Paula Clifford, *Angels with Trumpets*.
[28] Richard Bauckham, *Bible and Ecology*.
[29] Nick Spencer & Robert White, *Christianity, Climate Change and Sustainable Living*.
[30] Jonathan Moo & Robert White, *Hope in an Age of Despair*.
[31] George Marshal, *Don't Even Think About It*.
[32] Naomi Klein, *This Change Everything*.
[33] https://www.vatican.va/content/francesco/en/encyclicals/documents/papa-francesco_20150524_enciclica-laudato-si.html Accessed 9th Sept. 2021.
[34] https://www.anglicancommunion.org/mission/marks-of-mission.aspx Accessed 9th Sept. 2021.
[35] https://ecochurch.arocha.org.uk/ Accessed 9th Sept. 2021.
[36] https://www.globalcitizen.org/en/content/attenborough-effect-plastics/ Accessed 9th Sept. 2021.
[37] https://www.youtube.com/watch?v=BdIQlK6JGJs Accessed 9th Sept. 2021.

APPENDIX 3

Voices on Technology

As we discern what it means for each of us to be fully alive in our contemporary contexts, we have a rich vein of experience which might be mined in the form of all those working in a technological field. Followers of Jesus working in those fields will have insights which might be of particular value. If we are exploring the subjects of this book with people in local churches and other contexts, we may benefit from asking professionals working in these areas to help us as we reflect.

Between 2019 and 2020, I carried out research drawing upon the experience of over three hundred Christians working professionally in the fields of science, medicine or technology. These were the ten lessons which emerged from that work:

1. Christians are working across the breadth of technological, scientific and medical frontlines.

2. Christians can articulate the awesome and inspiring value of their work for society.

3. Christians are experiencing complex ethical challenges raised by their work.

4. Christians are open to having the wider church pray for their technological workplace.

5. Christians can help give our wider society glimpses into our possible technological future.

6. We, the wider church, have an untapped resource in our midst.

7. We, the wider church, have an opportunity to develop our prayer-life.

8. We, the wider church, have an opportunity to teach into these areas.

9. We, the wider church, have an opportunity to equip ethical decision-making in society.

10. We, the wider church, have an opportunity to further engage with these issues.

The full report is currently available online, titled *Voices on Technology*, via the websites of *Equipping Christian Leadership in an Age of Science* (https://www.eclasproject.org/new-research-offers-ten-lessons-on-experiences-of-christians-in-stem-professions/ Accessed 9th Sept. 2021) and *Christians in Science* (https://cis.org.uk/wp-content/uploads/2020/05/2020-05-Rev-Dr-J-Tomkins.pdf Accessed 9th Sept. 2021).

BIBLIOGRAPHY

Baker, S. (2011) *Final Jeopardy: Man vs. Machine and the Quest to Know Everything* (New York: Houghton Mifflin Harcourt).

Bauckham, R. (2010) *Bible and Ecology: Rediscovering the Community of Creation* (London: Darton, Longman & Todd Ltd).

Bostrom, N. (2005) *The Fable of the Dragon-Tyrant.* Available from: www.bostrom.org [Accessed 21 March 2011].

Bostrom, N. & Cirkovic, M. M. Editors (2008) *Global Catastrophic Risks* (Oxford: Oxford University Press).

Bostrom, N. (2014) *Superintelligence: Paths, Dangers, Strategies* (Oxford: Oxford University Press).

Cameron, N. M. de S. (2017) *Will Robots take your Job?* (Cambridge, UK: Polity).

Chivers, T. (2019) *The AI does not hate you: Superintelligence, Rationality and the Race to Save the World* (London: Weidenfeld & Nicolson).

Christian, B. (2011) *The Most Human Human: A Defence of Humanity in the Age of the Computer* (London: Viking).

Clifford, P. (2009) *Angels with Trumpets: The church in a time of global warming* (London: Darton, Longman & Todd Ltd).

Cole-Turner, R. Editor (2008) *Design and Destiny: Jewish and Christian Perspectives on Human Germline Modification* (Cambridge, MA: The MIT Press).

Collins, F. (2007) *The Language of God: A Scientist presents evidence for belief* (London: Pocket Books).

Collins, F. (2010) *The Language of Life: DNA and the Revolution in Personalised Medicine* (London: Profile Books).

Cooper, J. W. (1989) *Body, Soul & Life Everlasting: Biblical Anthropology and the Monism-Dualism Debate* (Grand Rapids, MI: William B. Eerdmans Publishing Company).

Davies, A. (2021) *Driven: The Race to Create the Autonomous Car* (New York: Simon & Schuster).

de Grey, A. with Rae, M. (2007) *Ending Aging: The Rejuvenation Breakthroughs that could reverse Human Aging in our lifetime* (New York: St Martin's Press).

Ellul, J. (1964) *The Technological Society* (Translated from the French by John Wilkinson) (Toronto: Vintage).

Ford, M. (2015) *The Rise of the Robots: Technology and the Threat of Mass Unemployment* (London: Oneworld).

Ford, M. (2018) *Architects of Intelligence: The Truth about AI from the people building it* (Birmingham, UK: Packt).

Fry, H. (2018) *Hello World: How to be Human in the Age of the Machine* (London: Doubleday).

Garreau, J. (2005) *Radical Evolution: The Promise and Peril of Enhancing Our Minds, Our Bodies – and What it Means to Be Human* (New York: Broadway Books).

Gay, C. M. (2018) *Modern Technology and the Human Future: A Christian Appraisal* (Downers Grove, IL: IVP Academic).

Giglio, L. (2012) *Symphony: I Lift my Hands* (Passion Talk Series DVD).

Greenfield, S. (2003) *Tomorrow's People: How 21st-Century Technology is Changing the Way We Think and Feel* (London: Allen Lane).

Harari, Y. N. (2011) *Sapiens: A Brief History of Humankind* (London: Vintage).

Harari, Y. N. (2015) *Homo Deus: A Brief History of Tomorrow* (London: Vintage).

Harari, Y. N. (2018) *21 Lessons for the 21st Century* (London: Jonathan Cape).

Hauerwas, S. (1981) *A Community of Character: Towards a Constructive Christian Social Ethic* (Notre Dame: University of Notre Dame Press).

Hill, K. (2017) *Left to their own Devices: Confident Parenting in a World of Screens* (Edinburgh: Muddy Pearl).

Houghton, J. (2004) *Global Warming: Third Edition* (Cambridge: Cambridge University Press).

Hutchings, D. & McLeish, T. (2017) *Let There be Science: Why God loves Science, and Science needs God* (Oxford: Lion).

Huxley, A. (1932) *Brave New World* (London: Vintage).

Jacobs, A. J. (2007) *The Year of Living Biblically: One Man's Humble Quest to Follow the Bible as Literally as Possible* (London: Arrow Books).

Joy, B. (2000) *Why the Future Doesn't Need Us. Wired* Issue 8.04 Available from: http://www.wired.com/wired/archive/8.04/joy.html [Accessed 17 May 2011].

Kasparov, G. (2017) *Deep Thinking: Where Machine Intelligence Ends* (London: John Murray).

Kelly, K. (2010) *What Technology Wants* (London: Viking).

Klein, N. (2014*) This Changes Everything: Capitalism vs. the Climate* (London: Allen Lane).

Kurzweil, R. (1999) *The Age of Spiritual Machines: When Computers Exceed Human Intelligence* (New York: Penguin).

Kurzweil, R. (2005) *The Singularity is Near* (New York: Penguin).

Kurzweil, R. & Grossman, T. (2005) *Fantastic Voyage: Live Long Enough to Live Forever* (London: Rodale).

Kurzweil, R. & Grossman, T. (2009) *TRANSCEND: Nine Steps to Living Well Forever* (New York: Rodale).

Kurzweil, R. (2019) *Danielle: Chronicles of a Superheroine and How you can be a Danielle* (Monument, CO: WordFire Press).

Lewis, C. S. (1956) *The Last Battle* (London: Diamond Books).

Lewis, C. S. (1978) *The Abolition of Man* (London: Fount Paperbacks).

Lennox, J. C. (2009) *God's Undertaker: Has Science Buried God?* (Oxford: Lion).

Lennox, J. C. (2020) *2084: Artificial Intelligence and the Future of Humanity* (Grand Rapids, MI: Zondervan Reflective).

Marshall, G. (2015) *Don't Even Think About It: Why our brains are wired to ignore climate change* (New York: Bloomsbury).

Maushart, S. (2011) *The Winter of our Disconnect: How one family pulled the plug and lived to tell/text/tweet the tale* (London: Profile Books).

McGrath, A. (2015) *Inventing the Universe: Why we can't stop talking about science, faith and God* (London: Hodder & Stoughton).

McIntosh, A. (2008) *Hell and High Water: Climate Change, Hope and the Human Condition* (Edinburgh, Birlinn).

McKibben, B. (2003) *Enough: Genetic Engineering and The End of Human Nature* (London: Bloomsbury Publishing Plc).

McKibben, B. (2010) *Eaarth: Making a Life on a Tough New Planet* (New York: St Martin's Griffin).

Messer, N. Editor (2002) *Theological Issues in Bioethics* (London: Darton, Longman and Todd).

Messer, N. (2011) *Respecting Life: Theology and Bioethics* (London: SCM Press).

Middleton, J. R. & Walsh, B. J. (1995) *Truth is Stranger Than it Used to Be* (London: SPCK).

Middleton, J. R. (2005) *The Liberating Image: The Imago Dei in Genesis 1* (Grand Rapids MI: Brazos Press).

Moo, J. & White, R. (2013) *Hope in an Age of Despair: The gospel and the future of life on earth* (Nottingham, Inter-Varsity Press).

Moore, P. (2000) *Babel's Shadow – Genetic Technologies in a Fracturing Society* (Oxford: Lion Publishing plc).

O'Donovan, O. (1984) *Begotten or Made?* (Oxford: Clarendon Press).

Polkinghorne, J. (1991) *Reason and Reality* (London: SPCK).

Polkinghorne, J. (1994) *Quarks, Chaos & Christianity* (London: Triangle).

Ponsonby, S. (2009) *More: How you can have more of the Spirit when you already have everything in Christ* (Eastbourne: David. C. Cook).

Rees, M. (2003) *Our Final Century: Will Civilisation Survive the Twenty-First Century?* (London: Arrow Books).

Roose, K. (2021) *Futureproof: 9 Rules for Humans in the Age of Automation* (London: John Murray).

Sacks, J. (2011) *The Great Partnership: God, Science and the Search for Meaning* (London: Hodder & Stoughton).

Silver, L. M. (1999) *Remaking Eden: Cloning, Genetic Engineering and the Future of Humankind?* (London: Phoenix Giant).

Spencer, N. & White, R. (2007) *Christianity, Climate Change and Sustainable Living* (London: SPCK).

Susskind, R. & Susskind, D. (2015) *The Future of the Professions: How Technology will transform the work of Human Experts* (Oxford: Oxford University Press).

Tegmark, M. (2017) *Life 3.0: Being Human in the age of Artificial Intelligence* (London: Penguin).

Tomkins, J. (2013) *Better People or Enhanced Humans?: What it might mean to be fully alive in the context of Human Enhancement* (Chesham, UK: Sunnyside Books).

Torrance, T. F. (2008) *Incarnation: The Person and Life of Christ* (Downers Grove, IL: InterVarsity Press).

Turkle, S. (2017) *Alone Together: Why we Expect More from Technology and Less from Each Other* (New York: Basic Books).

Welby, J. (2016) *Dethroning Mammon: Making Money Serve Grace* (London: Bloomsbury).

Wells, S. (2004) *Improvisation: The Drama of Christian Ethics* (London: SPCK).

Williams, R. (2020) *Candles in the Dark: Faith, hope and love in a time of pandemic* (London: SPCK).

Wright, T. (2011) *Revelation for Everyone* (London: SPCK).

Wright, T. (2020) *God and the Pandemic: A Christian Reflection on the Coronavirus and its Aftermath* (London: SPCK).